Analytical Atomic Absorption Spectroscopy

SELECTED METHODS

Analytical Atomic Absorption Spectroscopy

SELECTED METHODS

Jon C. Van Loon

*Departments of Geology and Chemistry
and
The Institute for Environmental Studies
University of Toronto
Toronto, Ontario
Canada*

ACADEMIC PRESS 1980
A Subsidiary of Harcourt Brace Jovanovich, Publishers
New York   London   Toronto   Sydney   San Francisco

ACADEMIC PRESS, INC.
111 Fifth Avenue, New York, New York 10003

United Kingdom Edition published by
ACADEMIC PRESS, INC. (LONDON) LTD.
24/28 Oval Road, London NW1 7DX

Library of Congress Cataloging in Publication Data

Van Loon, Jon Clement, Date
    Analytical atomic absorption spectroscopy.

    Includes bibliographical references and index.
    1. Atomic absorption spectroscopy. I. Title.
QD96.A2V36   543'.0858   79–25448
ISBN 0–12–714050–6

PRINTED IN THE UNITED STATES OF AMERICA

83   9 8 7 6 5 4 3

To the loving thoughts I hold for my wife, Maureen, and for my children, Lisa, Melisa, and Jon

Contents

Preface

This book was conceived to appeal to the practicing analyst. Theory is presented in a very descriptive, nonrigorous fashion. Citations have been kept to a minimum to improve the flow of material and to avoid the impression that the reading of these references is essential. The general principles of atomic absorption spectroscopy are now well-established, and it seems desirable to cease the referencing of each point of theory.

The procedures were selected with a view to their reliability. Emphasis was placed on choosing thoroughly tested methods, preferably ones that were evaluated by using standard reference samples or inter-laboratory comparison. Unfortunately, this process and the personal bias of the author will inevitably have resulted in the exclusion of good procedures. The methods are given, as much as possible, in the words of the original author. This has been done to avoid misinterpretations, which can often negate the usefulness of a method.

There will inevitably be gaps in the chapters where the analyst might feel a procedure should have been given. A conscious attempt has been made to recognize the strengths and weakness of atomic absorption spectroscopy and thereby avoid including procedures for elements and applications for which the technique is poorly suited. Good examples of the latter are the omission of most major element analyses, analyses better done by flame emission and indirect methods. Other gaps will exist. In spite of the burgeoning atomic-absorption-methods literature, a critical review has shown that no reliable procedure has been published for many applications. Again, I am painfully aware of omissions that may occur due to the subjective nature of such a process.

Analytical methods have been grouped into chapters in a manner that minimizes overlap and repetition. However, there are instances where procedures useful for several sample types have been placed in the most suitable chapter and reference is then given in text to other applications.

In addition to the conventional chapters, a short section has been presented on elemental speciation using atomic absorption spectroscopy. This latter field, so crucial to the solution of environmental, biological, and health related problems, is in its infancy. Few tested procedures yet exist in this area. However, the study of elemental speciation is a very important new direction being taken by atomic absorption spectroscopy. Therefore, the available work has been included here, to introduce analysts to the power and simplicity of the approach.

Jon C. Van Loon

Acknowledgments

The present author is greatly indebted to the many researchers and publishers who kindly gave permission for their procedures to be reproduced in this volume and to Perkin-Elmer Corporation, Norwalk, Connecticut; Varian Tektron Pty. Ltd., Springvale, Australia; and CSIRO, Division of Chemical Physics, Melbourne, Australia for photographs and other materials for inclusion in the book.

1
General Principles

INTRODUCTION

The phenomenon of the absorption of radiation by atoms has been used for investigations in physics since the early part of the nineteenth century, when Frauenhöfer observed a number of dark lines in the sun's spectrum. The first analytical application of atomic absorption was to the determination of mercury by Müller (1). Not until 1955, when Walsh (2) discovered the general usefulness of the approach to elemental analysis, was real analytical atomic absorption spectroscopy born. In the relatively short period of two and one-half decades since this development, atomic absorption spectroscopy has become one of the most important techniques for the analysis of the elements.

During the early commercialization of atomic absorption spectroscopy, extravagant proclamations were made concerning the general lack of interferences encountered with this technique. This is peculiar in that Walsh, in his original paper, limited his claims for superiority to a simpler pattern of atomic spectral interference and a greater tolerance to thermal fluctuations in the atomizer compared to emission spectroscopy. Another rather amusing characteristic of early research in this field was the great deal of wasted time spent by many workers, including the present author, in rediscovering interferences in flames which had been known to emission spectroscopists for up to 50 years.

Emission spectra are usually produced thermally using high-energy sources such as high-temperature flames, arcs, or sparks, and they are usually very complex. For a given element only a small fraction of the atoms obtained is in an excited state and this fraction is highly temperature dependent. Hence, small temperature fluctuations cause appreciable variations in emission. The ground-state atom population, important in absorption, is much less affected by small temperature fluctuations. Many atomic species generated in high-energy thermal sources emit radiation. This results in a complex spectrum that must be resolved using an expensive, high-resolution spectrometer.

Absorption spectra are relatively simple, requiring the use of a low-priced monochromator. The likelihood of atomic spectral overlap is high with emission spectroscopy, whereas this possibility is much reduced with atomic absorption spectroscopy.

Few techniques of chemical analysis have the inherent simplicity of atomic absorption spectroscopy. Thus, the technique is commonly used by those with limited knowledge of analytical chemistry or atomic spectroscopy. It is important to emphasize the potential consequences of this fact. The present author is frequently confronted with manuscripts and theses written by researchers in biological, health, geological, and metallurgical sciences that are based on results obtained by atomic absorption spectroscopy. In many cases it is impossible to judge the validity of the data presented. These workers seldom give sufficient procedural detail to indicate whether precautions have been taken to avoid common sources of interference (e.g., nonspecific absorption). Standard reference materials are seldom analyzed.

No technique of chemical analysis is universally applicable. Figure 1 is a periodic table showing only those elements that can be analyzed directly by atomic absorption. Of those shown, osmium, wolfram, zirconium, hafnium, niobium, tantalum, and the rare earths are better

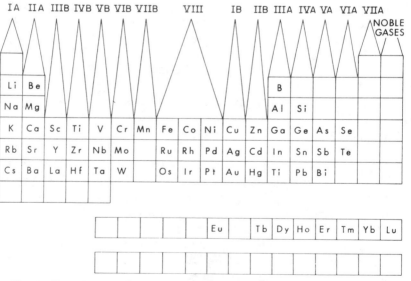

Fig. 1. Periodic table showing elements determinable by atomic absorption.

handled by other techniques. These refractory elements present difficulties and exhibit poor sensitivity in flame atomizers. Alkali metals are easily handled by flame emission and many workers still prefer this approach. Below the parts per million (ppm) level, neutron activation and related nuclear techniques, when available, must be considered competitive. In the percentage range, x-ray fluorescence spectroscopy is superior. There is a tendency among analysts to force atomic absorption spectroscopy to do analyses for which it is poorly suited. The meteoric rise in popularity of atomic absorption must not be allowed to obscure the niche it best fills among other analytical techniques.

ATOMIC SPECTROSCOPY

Techniques commonly referred to as atomic spectroscopy usually include atomic emission, atomic absorption, and atomic fluorescence spectroscopy. These involve valence electron transitions yielding radiation with wavelengths in the ultraviolet–visible region of the spectrum. X-Ray fluorescence and neutron activation, although atomic techniques, are not usually referred to as atomic spectroscopy. An abundance of analytical procedures exist that involve either atomic emission or absorption spectroscopy, but little analytical use has been made of atomic fluorescence spectroscopy.

Atomic Spectra

Electron orbits in an atom are characterized by the major and azimuthal quantum numbers n and l, respectively. When an electron undergoes a transition from a higher energy level (E_{nl}) to a lower energy level $(E_{n_1 l_1})$, light of frequency

$$\nu = (E_{nl} - E_{n_1 l_1})/h = \Delta E/h$$

is given off. In terms of wavelength,

$$\lambda = c/\nu = hc/\Delta E$$

The constants h and c in these questions are Planck's constant and the velocity of light, respectively. Thus, electronic transitions can be discussed in terms of frequency ν, energy E, and wavelength λ. The latter is most frequently used in atomic absorption spectroscopy. The parameters ΔE, ν, and λ have unique values for a given electronic transition. An element can undergo many electronic transitions. This results in a

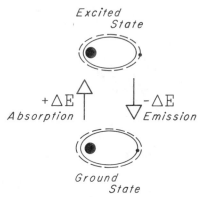

Fig. 2.   Relationship between emission and absorption of energy by a valence electron.

series of sharp lines—a spectrum—which is uniquely characteristic of each element.

Emission and absorption can be related as shown in Fig. 2. An electron of an atom in its lowest energy state, the ground state, can absorb a quantum of energy $(+\Delta E)$ and undergo transition to a low-lying excited state. Emission occurs when this quantum of energy is released $(-\Delta E)$ and the electron returns to the ground state. A transition to and from the ground state, as illustrated in this example, is called a resonance transition. Resonance lines are the most useful analytical lines for atomic absorption spectrometry.

An energy level graphical representation of an atom is usually given by a Grotian diagram (3). A Grotian diagram for sodium is shown in Fig. 3. Sodium has one valence electron. When this electron is in the ground state (3s), the energy-of-transition-scale (vertical, electron volts) shows zero. Resonance transitions emanate from this energy level, with the most commonly used analytical line for absorption measurements being 5889.9 Å. This transition from the $3s_{1/2}^2$ to the $3p_{3/2}^2$ energy level is the lowest energy transition for sodium and represents an energy gain by the electron of 2.2 eV.

The horizontal axis of the Grotian diagram shows the term symbol. This shows that p levels have two terms, $3p_{3/2}^2$ and $3p_{1/2}^2$. The detailed discussion of this aspect of the Grotian diagram is beyond the scope of this book. However, it should be pointed out that this indicates that transitions from p levels for one-valence-electron elements result in doublets. In the case of sodium, the 5895.9 Å transition represents only a slightly different energy than the 5889.9 Å line and these two are the well-known D line doublets of the sodium spectrum.

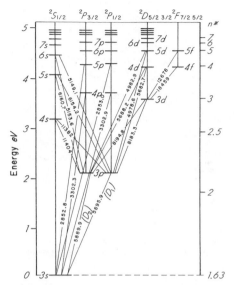

Fig. 3. Grotian diagram (3) for sodium.

Electronic transitions in an atom are governed by selection rules that designate which possible transitions are permitted and forbidden. For a good discussion of this and other more fundamental aspects of spectroscopy, the reader is referred to Mitchell and Zemansky (4) and Herzberg (5).

ABSORPTION EXPRESSION

An Absorption Line

When a parallel beam of continuous radiation of intensity I_0 passes through a cell containing atomic species of an element, the transmitted radiation I_ν will show a frequency distribution as given in Fig. 4 (4). The atomic species is said to possess an absorption line at frequency ν_0, where ν_0 is the frequency at the center of the line. The absorption coefficient of the atomic vapor k_ν is defined by

$$I_\nu = I_0 e^{-k_\nu b}$$

where b is the thickness of the absorbing layer, and the integrated absorption

$$\int k_\nu \, d\nu = \frac{\pi e^2}{mc} N_\nu f$$

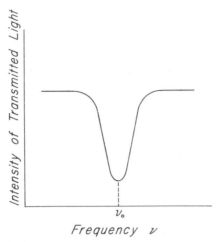

Fig. 4.   Frequency dependence of transmitted radiation.

where N_ν is the number of atoms per cubic centimeter that can absorb in the frequency range ν to $\nu + d\nu$, f is the oscillator strength (the average number of electrons per atom that can be excited by the incident radiation), e is the charge on an electron, c the velocity of light, and m the mass of an electron. Tables of oscillator strength are available to allow a comparison of transition probabilities for a given line and a given element. Figure 5 (4) depicts the variation in k_ν with frequency. The maximum value of the absorption coefficient is k_{max} and its value at half-height is designated $k_{max}/2$.

Relation between Emission and Absorption and Linewidths

Natural linewidths are of the order of 10^{-4} Å. Broadening occurs due to self-absorption, the Doppler effect, and collisional processes. In the majority of cases the most important collisional broadening is due to collisions of the analyte atoms with foreign gases (Lorentz broadening). Line sources such as hollow-cathode lamps are generally used in atomic absorption spectroscopy. Hollow-cathode lamp emission occurs in an inert gas atmosphere at a reduced pressure of about 1 torr. Flames or furnaces are used as absorption cells in atomic absorption spectroscopy. In contrast to line sources, atoms in flames or furnaces are at or near ambient pressures and much collisional broadening occurs. Thus the emission linewidth from a hollow-cathode lamp is narrow compared to the absorption line. With reference to Fig. 5, the width of an absorption

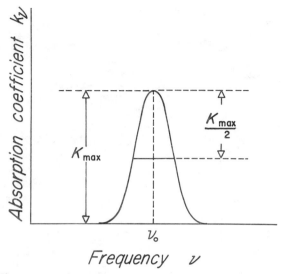

Fig. 5. Variation of absorption coefficient with frequency.

line is taken to be the frequency interval at $k_{max}/2$. It is therefore possible in conventional atomic absorption analysis to measure absorption at the center of the absorption line.

Relation between Absorption and Concentration

Absorption of monochromatic radiation is governed by the following two laws:

Lambert's law. Light absorbed in a transparent absorption cell is independent of incident light intensity. An equal fraction of the light is absorbed by each successive layer of absorbing medium.

Beer's law. Absorption of light is likewise exponentially proportional to the number of absorbing species in the path of the light beam.

Referring to Fig. 6, the incident beam of monochromatic radiation I_0 falls on an absorption cell of length b. The transmittance is given by $T = e^{-kbc}$. Then since

$$\log_{10}(1/T) = \log_{10}(I_0/I) = abc$$

and

$$\log_{10}(I_0/I) = A$$

where A is the absorbance, then $A = abc$, where a is a constant for a

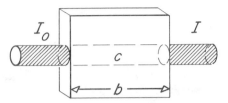

Fig. 6. Atomic absorption cell of length b.

given system and c is the concentration of the analyte atoms in the flame.

This expression, known as the Beer–Lambert law, predicts a linear relationship between absorbance and concentration as long as a and b remain constant. The linear relationship is between the concentration of analyte atoms in the atomizer and not necessarily analyte in the sample solution. To obtain the latter it is necessary to carefully avoid interference problems outlined in subsequent sections.

It is important to note that the atomic absorption instrument must compute $\log (I_0/I)$, that is, $\log I_0 - \log I$ rather than a small emitted signal as is the case in emission spectroscopy. Failure to appreciate that different parameters, with no direct relationship between them, are being measured has led many workers to claim a fundamental improvement in sensitivity of absorption over emission spectroscopy.

Figure 7 shows a linear calibration graph. Three solutions of known concentration 1, 2, and 4 $\mu g/ml$ of an element are nebulized into an atomic absorption flame and the absorbances are found to be A_1, A_2, and A_3, respectively. As can be seen, this yields a straight line as

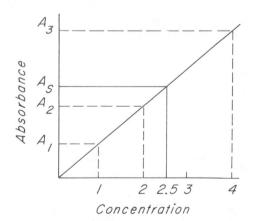

Fig. 7. Linear calibration graph.

predicted by the Beer–Lambert law. A sample to be analyzed is then nebulized. The absorbance obtained is A_s. By interpolation on the calibration graph a concentration of 2.5 μg/ml is obtained for the sample.

It is common to obtain calibration graphs with departures from ideality. The curves in Fig. 8 summarize frequently encountered problems. Curve A does not pass through zero absorbance-concentration. This problem, due to nonspecific absorption, is discussed in detail on p. 39.

Nonlinearity in Analytical Calibration Graphs

The concentration working range for an element determined by atomic absorption spectroscopy is generally 4 to 5 orders of magnitude. At the upper end of the concentration range, it is common to have the graph bend toward the concentration axis (Fig. 8, curve B). Less frequently, the calibration graph curves throughout the whole working range (curve D).

The Beer–Lambert law is valid for monochromatic radiation. Reasons for the bending of curves have been postulated by a number of researchers. de Galan and Samaey (6) indicate two common causes of this problem. Both result from failure of the monochromator and slit system to prevent multiple, close-spaced lines from reaching the detector. If several lines of differing absorption coefficient (e.g., unresolved multiplets or nonabsorbing lines) fall on the detector, a nonlinear relationship will result between absorbance and analyte concentration. The calibra-

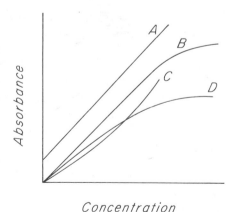

Concentration

Fig. 8.   Calibration graph showing departures from ideality.

tion curve will bend toward the concentration axis (Fig, 8, curves B and D).

The linewidth of the emission source must be narrower than the absorption linewidth. If this is not the case, a nonlinear relation between absorbance and concentration will occur (curve D) because of the variation in k_ν with frequency shown in Fig. 5. The greatest cause of line broadening, in good hollow-cathode lamps, is the use of excessive lamp currents.

The following procedures should be followed to maximize chances of obtaining good calibration curve linearity:

(1) Use the manufacturers recommended current or power settings for light sources. Higher values can result in serious line broadening.

(2) For elements (e.g., nickel, 2320 Å) that have a number of other lines closely spaced around the chosen analytical line, use as narrow a slit width as is consistent with an acceptable signal to noise (S/N) ratio. In most cases the manufacturer's recommendation should be followed.

(3) Great care should be taken to ensure that measurements are made at the very peak of the chosen analytical line.

Ionization results in a curve that bends toward the absorbance axis (Fig. 8, curve C). This is due to the concentration dependence of the ratio of ionized to un-ionized atoms. This ratio decreases at higher concentrations.

AN ATOMIC ABSORPTION SPECTROMETER

The essential components of an atomic absorption spectrometer are diagrammed in Fig. 9. In most commercial instrumentation, A is a hollow-cathode lamp, B a flame or electrothermal device, C a grating monochromator, and D a photomultiplier. The first atomic absorption spectrometer exhibited by Walsh (7) is shown in Fig. 10. This equipment is remarkably similar in principle to modern atomic absorption units. In fact, with the exception of electrothermal atomizers and devices for background correction, little of fundamental importance has been introduced into commercial units to this day.

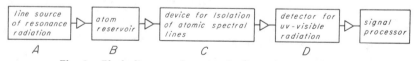

Fig. 9.   Block diagram of an atomic absorption spectrometer.

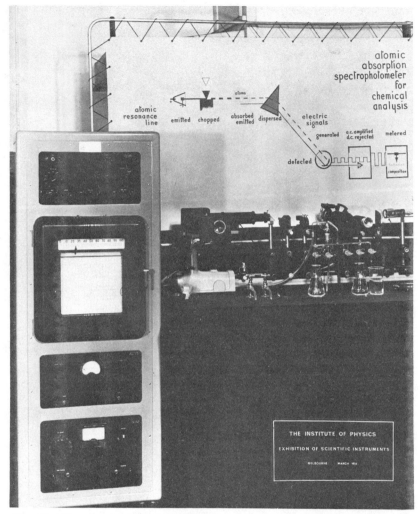

Fig. 10. First atomic absorption spectrometer exhibited by Walsh (7), CSIRO Division of Chemical Physics, Melbourne, Australia.

The principle of operation of atomic absorption instrumentation is simple. For most instrumentation this can be described as follows.

The hollow-cathode lamp emits radiation characteristic of the cathode material, usually a single element (analyte). This beam, consisting largely of resonance radiation, is electronically or mechanically pulsed. Analyte atoms are produced thermally in the atom reservoir. Ground-state atoms, which predominate under the experimental conditions, absorb resonance radiation from the lamp, reducing the intensity of the incident beam. The monochromator isolates the desired resonance line and allows this radiation to fall on the photomultiplier. An electrical signal is generated. The electronics of the unit are designed to respond selectively to the pulsed radiation emanating from the radiation source. Signal processing occurs, which results in electronic output proportional to the absorption by the analyte atoms.

Absorption spectra are relatively simple, consisting predominantly of resonance transitions. The monochromator need not be capable of especially high resolution, its function being to isolate the resonance line of interest and to diminish the light flux from nonabsorbing lines in the source and emission from the flame. Because the signal processor is tuned to respond only to the pulsed radiation from the lamp, emission from the flame does not produce signal but does produce noise. The very small fluorescence signal from analyte atoms is emitted isotropically and will be an insignificant component of the signal falling on the photomultiplier.

Instrument offerings to date are for single- or two-element analysis. Development of simultaneous multielement atomic absorption equipment has been hindered by problems with optical alignment of components, the need for different operating conditions for many elements, and the relatively short linear range of calibration curves.

RADIATION SOURCES

Hollow-Cathode Lamps

Hollow-cathode atomic spectral lamps are the most common radiation sources for atomic absorption spectroscopy. These lamps can produce resonance radiation of narrow linewidth, typically < 0.01 Å, for most elements that are determinable by atomic absorption. It is important in atomic absorption that the linewidth of the radiation source be narrower than that of the absorption profile. Failure to have this condition results in poorer sensitivity and badly curved calibration graphs. To maximize the probability of obtaining radiation of narrow linewidth, the manufacturers' instrumentations pertaining to operating currents should be closely followed.

Figure 11 shows the essential components of a hollow-cathode lamp. The lamp is evacuated and filled with about 2 torr of either argon or neon. A small current is passed between the anode and the cathode,

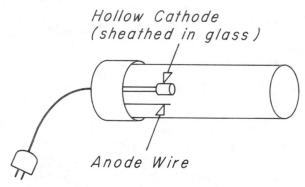

Fig. 11.   Hollow-cathode lamp.

resulting in ionization of inert-gas atoms. Atoms of the cathode metal (analyte) are sputtered from the surface due to interactions with these ions. Resonance radiation results when the ground-state atoms are excited and then decay back to the ground state. Although the discharge may appear to the eye to be hot, the cathode temperature is usually only 300°–400°C. Excitation results from collisions between analyte atoms and inert-gas ions in the discharge.

The cathode is constructed from the metal or an alloy of the element being determined. The spectrum of a good lamp consists mainly of spectral lines of the element of interest. The latter will be both resonance and nonresonance lines. Lines will also result from the filler gas. If hollow-cathode radiation other than the desired resonance line (e.g., nonabsorbing lines) is detected by the photomultiplier, poor sensitivity and curved calibration graphs are obtained. A commonly encountered example is the determination of nickel using the 2320 Å line, where the nickel ion line at 2316 Å may cause a problem.

Hollow cathodes have a life expectancy of 1 to 2 years, depending on the element, amount of use, average current employed, and problems due to gas leakage into the lamp. Shelf storage of lamps for long periods has an adverse effect on expected lifetime. Each hollow cathode should be run for a period of 30 min monthly, whether they are required for analysis or not. Prolonged use of lamp currents at or above the recommended maximum drastically reduce lamp lifetime.

Multielemental hollow-cathode lamps are available from some suppliers. Considering the cost of single-element lamps (currently $150–250) and the need for one lamp for each element being determined, multielement lamps have an obvious attraction. However, it is the present author's recommendation that multielement lamps not be used. This is because atomic spectral interferences have been reported for some elements using these sources. In addition, multielement lamps do not age well, often showing marked decreases in emission intensity for some of the constituent elements.

Electrodeless Discharge Lamps

It is difficult to make stable, long-lived, and bright hollow-cathode lamps for a few elements. The latter include arsenic, selenium, tellurium, tin, and lead. Thus, several companies, principally Perkin-Elmer and Westinghouse, now offer electrodeless discharge lamps (see Fig. 12) for these and a number of other elements. Prior to the commercial offerings, electrodeless discharge lamps had been manufactured in research laboratories and used mainly for atomic fluorescence spectroscopy. Laboratory-constructed lamps were characterized by poor stability, reliability, and reproducibility from one lamp to the next.

Electrodeless discharge lamps are constructed by sealing a small amount of the pure metal or pure metal salt, usually the iodide, of the element of interest into a silica bulb together with an inert gas at low pressure. This bulb is placed in the cavity of a radio frequency (RF) coil powered at 27 MHz. The components are then housed in a lamp body of size similar to a hollow-cathode lamp.

A separate power supply is necessary for electrodeless discharge lamps. To operate these lamps, the supply is turned on and adjusted to the recommended power setting. A warm-up period of at least 30 min is necessary to obtain stable lamp output. No greater than the recommended lamp power should be used. At this setting, maximum sensitivity is achieved. When run above the recommended power, electrodeless discharge lamps yield a more intense spectral output. However, line broadening, an undesirable effect in atomic absorption work, may also occur. Line broadening results in poorer atomic absorption sensitivity.

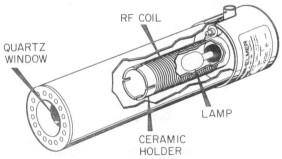

Fig. 12.   Electrodeless discharge lamp (Perkin-Elmer Corp.).

When run according to the manufacturer's instructions, electrodeless discharge lamps have a much longer life span than hollow-cathode lamps. They are to be highly recommended, especially for the above-mentioned elements.

Other Atomic Spectral Radiation Sources

Vapor Discharge Lamps

Sodium and potassium hollow cathode lamps are often unsatisfactory. Vapor discharge lamps are available for these elements and are recommended if they can be used with the instrumentation in question. Again, it is important to use only the recommended current with these lamps.

High-Intensity Hollow-Cathode Lamps

Ordinary hollow-cathode lamps have barely sufficient output for several elements. High-intensity lamps may be constructed by placing a second pair of electrodes in a lamp so that the atom cloud produced by the primary discharge is bombarded by a high-current discharge produced by these electrodes. A significant gain in spectral output is thus obtained. None of these lamps is presently commercially available. However, should this approach be applicable to the refractory elements (satisfactory electrodeless discharge lamps of most refractory elements seem unlikely), and preliminary research in North America and Australia suggests this may be the case, then commercial lamps would likely be offered.

ATOMIZATION SYSTEMS

In order to have atomic absorption, it is necessary to produce free ground-state atoms of the element of interest. The production of free atoms occurs in an atomizer and is called atomization. A variety of commercial atomizers are available for use with atomic absorption equipment, the most common being flames and furnaces.

Burner–Nebulizer Systems

There is only one type of burner in general use today for atomic absorption spectroscopy—the laminar flow premix system. "Total" consumption burners, used in flame emission, are not well suited for absorption work.

Nebulizer

Liquid sample is introduced into a burner through the nebulizer by the venturi action of the nebulizer oxidant. Figure 13 shows a cross section of a nebulizer–premix chamber–burner assembly. In passage through the nebulizer the liquid stream is broken into a droplet spray. During nebulization some liquids are broken into a finer mist than others. For example, an organic solvent such as methyl isobutyl ketone is more efficiently converted into a fine droplet size than is water. This is due to reduced surface tension in the case of the former. Since the amount of fine mist per unit sample volume reaching the flame affects the signal magnitude, it is important to nebulize samples and standards of similar solvent composition.

The nebulizer draws solution up a tube of narrow diameter (capillary). High-viscosity fluids flow through capillaries at a slower rate than low-viscosity fluids. Hence it is important to keep the viscos-

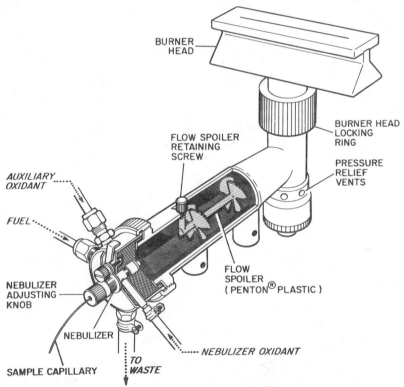

Fig. 13. Burner nebulizer system (Perkin-Elmer Model 305B).

ity of sample and standard solutions similar to avoid the possibility of physical interference problems (see p. 42).

Nebulizer capillaries present two other serious problems. They readily become clogged by particulate material and they sometimes corrode. It is very important to keep particulate matter out of nebulizers even if this requires a time-consuming filtration. The author has had the annoying experience of requiring the replacement of an expensive nebulizer because of plugging by an immovable angular silica fragment.

Most standard nebulizers will not withstand severe acid attack, e.g., aqua regia. Manufacturers sell, as an accessory, acid-resistant nebulizers. The author recommends the purchase of such a unit.

The flow rate of many nebulizers can be easily adjusted. A solution uptake rate of 4 to 6 ml/min is commonly employed for water solutions. When combustible organic solvents are utilized the nebulizer flow rate must be slowed.

Total Consumption Burners

This terminology is used to describe a burner where all the liquid that enters the nebulizer passes through into the flame. The term "total consumption" is a misnomer since much of the liquid spray produced in the nebulizer is not evaporated but passes through relatively untouched. So-called total consumption burners are, in fact, very inefficient in the production of atoms because of the short residence time of a unit of sample in the flame and the large droplet size of the aerosol. Because of this and because the optical path length of the source radiation in the flame is short, total consumption burners are seldom used in atomic absorption analysis.

Premix Burners

A large variety of premix burners are commercially available. These burners contain, in addition to a nebulizer, a premix chamber and a burner head (Fig. 13). Premix burners can be used with a wide variety of fuels and oxidants, but must not be employed for high-burning-velocity mixtures such as oxygen–acetylene.

Premix chambers are designed to mix the fuel, oxidant, and sample. Contrary to the case of the total consumption burner, not all the liquid entering a premix burner passes into the flame. The premix chamber allows large droplets, which are deleterious to the absorption process, to condense (aided by the flow spoiler in the system depicted in Fig. 13) and pass out of the chamber through the liquid drain tube. Rejection of these large drops helps minimize light-scattering effects in the flame.

As indicated above, the solvent type influences the fraction of fine droplets produced. For some solvents as much as 95% of the sample is rejected.

Burner heads for premix burners come in many designs. Most contain a single slot 5–10 cm in length. These relatively long slots allow the radiation beam from the source to traverse a lengthy path of atoms. The magnitude of the absorbance is related to the path length of the radiation beam in the flame.

The length of the slot, however, is restricted by flashback problems and turbulent flame characteristics (noise) to a maximum of about 10 cm. Some workers have attempted to increase the effective slot length of the burner by passing the light beam through the flame length several times using mirrors. This approach fails because the increase in noise is directly proportional to the number of flame–air interfaces through which the beam of radiation passes. Also some light is lost with each reflection.

Boling (8) designed a three-slot burner head that accommodates the full width of the radiation beam (i.e., fills the horizontal optical aperture) and hence enhances absorption. A single slot of the required width would not support a flame and flashback occurs. The three-slot Boling burner is excellent for some elements (e.g., chromium) and has been standard on some equipment. However, there have been conflicting reports of more complex interference patterns obtained with this burner and the worker is cautioned accordingly.

Nitrous oxide–acetylene burner heads must be constructed from high-temperature alloys. Slots generally do not exceed 5 cm in length. It is essential to use only designated nitrous oxide burner heads to minimize the possibility of a serious explosion with this oxidant.

Some manufacturers offer a high solids burner. This is a burner head designed to prevent plugging by the accumulation of precipitated solids on the slot. Such burners typically have a slightly wider slot width. Many workers will find this a useful accessory.

There is some reference in the literature to modified burner systems such as the heated premix chamber and the Fuwa tube (9). These devices although useful in some applications have not become generally available.

Flames

To date nothing has appeared to challenge the flame as the most generally useful atomizer for atomic absorption spectroscopy. This statement may be made despite the number of recent developments in electrothermal devices. Much more work is essential, despite numerous claims to the contrary by the manufacturers, before electrothermal devices are useful for a wide range of applications.

The least-understood component of atomic absorption equipment is the flame. Despite volumes of research, quantitative flame reactions have yet to be elucidated for most flame types. However, useful descriptive terminologies have been developed.

Flames are not uniform in composition, length, or cross section. Figure 14 shows the three zones that can be identified in flames burning on a premix burner, the primary combustion, interconal, and secondary combustion zones.

The primary combustion zone, also termed the reaction zone or inner cone, is only rarely used for absorption work. Emission is intense. Excessive radiation falling on the photomultiplier can cause noise problems (shot noise). Conditions are very reducing. Thermodynamic equilibrium is not obtained in this part of the flame.

The interconal zone separates the primary and secondary combustion zones. Thermodynamic equilibrium is obtained, making this area very useful for absorption measurements.

In the secondary combustion zone the main reaction is

$$2CO + O_2 \rightarrow 2CO_2$$

This area of the flame is cooler and more oxidizing than the interconal zone. Atomic absorption is seldom done in this area of the flame.

The reader may see reference in the literature to shielded flames. Shielding a flame with a sheath of inert gas, blown around the outside of the flame, causes an elongation of the interconal zone. Shot noise problems can thus be kept to a minimum. However, shot noise is

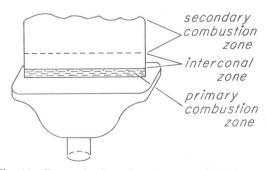

secondary combustion zone

interconal zone

primary combustion zone

Fig. 14.   Zones of a flame burning on a premix burner.

seldom a problem in atomic absorption. Hence, combined with the need for inert gases and burner modifications, this makes the use of shielded flames in routine absorption work undesirable. Shielded flames are very useful for the related technique of nondispersive atomic fluorescence spectroscopy.

Atomization Process in Flames

In this volume, atomization relates to the formation of atoms and not to the production of a fine spray as the word is sometimes taken to mean. The latter function is performed by the nebulizer. Processes in the flame are illustrated in Fig. 15.

The liquid mist produced by the nebulizer and further processed to extract the finest droplets in the premix chamber evaporates on entering the flame. The compounds produced pass upward into hotter regions of the flame. These compounds, agglomerated into small solid particles, begin to vaporize as their vapor pressures become appreciable. As the molecules pass into the hotter regions of the flame, they dissociate into atoms.

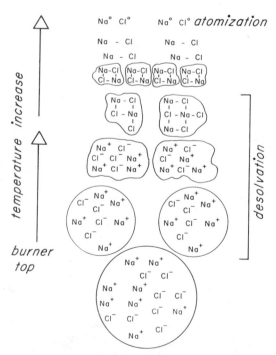

Fig. 15. Atomization processes resulting when liquid mist is introduced into a flame.

The preceding description is, of course, the ideal case. For a number of analyte elements, refractory compounds are produced that only fractionally break down at ordinary flame temperatures. This results in a poor atomization efficiency. For these elements a high-temperature nitrous oxide–acetylene flame is preferable.

Easily ionizable atoms may, in a step subsequent to atomization, lose one or more electrons, particularly if very hot flames are used. If other easily ionizable elements are present an ionization interference is experienced.

Flame Types

The most commonly employed flame in atomic absorption is the air–acetylene flame. Argon entrained air–hydrogen and nitrous oxide–acetylene flames are also used. Table I is a compilation of flame detection limits (air–acetylene flame unless otherwise noted) for solutions low in dissolved salts (10). Choice of flame type is governed by the temperature required. From the point of view of efficient vaporization and atomization it is best to use a high-temperature flame. This will minimize problems due to chemical and nonspecific background interferences. However, hotter flames, e.g., nitrous oxide–acetylene (temperature about 3000°C) cause severe ionization problems. In addition, hotter flames tend to produce intense emission, which in turn creates shot noise problems. The latter causes a degradation in precision of measurements and inferior detection limits.

In the region below 2000 Å, flame gas absorption of the source radiation can be a very serious problem. The argon-entrained air–hydrogen flame produces the best flame transparency conditions. This flame is often employed for arsenic (1937 Å) and selenium (1961 Å) determinations.

Below 1900 Å the absorption of the source signal due to air is so strong that determinations with lines of wavelengths shorter than this are impossible without using an inert-gas-purged or vacuum spectrometer. Unfortunately, important elements such as phosphorous, nitrogen, sulfur, and mercury have resonance lines in the vacuum ultraviolet. Use of an argon–hydrogen flame, temperature less than 2000°C, results in serious chemical interference problems which may obviate its use for complex samples.

When working in the wavelength range above 2000 Å and when chemical interferences are minimal, or can be overcome easily with releasing agents, the air–acetylene flame (temperature about 2300°C) is convenient and safe.

Table I

Flame Detection Limits (μg/ml) for a Perkin-Elmer 603 Using Conditions Recommended by the Manufacturer (10)

Element	Detection limit[d]	Element	Detection limit[d]	Element	Detection limit[d]	Element	Detection limit[d]
Ag	0.002	Co	0.01	Mg	0.0001	Ru	0.07
Al[a]	0.02	Cr	0.003	Mn	0.002	Sb	0.04
As	0.2	Cu	0.002	Mo[a]	0.02	Se[c]	0.20
Au	0.01	Fe	0.005	Na	0.0002	Si[a]	0.02
B[a]	0.7	Ga[a]	0.05	Ni	0.005	Sn[a,c]	0.07
Ba[a,b]	0.008	Ge[a,c]	0.10	Os[a]	0.08	Sr[b]	0.002
Be[a]	0.001	Hg[c]	0.25	Pb[c]	0.01	Te[c]	0.03
Bi	0.025	Ir	0.6	Pd	0.02	Ti[a]	0.04
Ca	0.0005	K	0.002	Pt	0.05	Tl[c]	0.01
Cd[c]	0.001	Li	0.0003	Rh	0.004	V[a]	0.04
						Zn	0.001

[a] Nitrous oxide–acetylene flame.
[b] 1000 μg/ml K added to suppress ionization.
[c] Electrodeless discharge lamp.
[d] Detection limit = standard concentration × standard deviation/mean.

Flame Conditions

In the preceding section, flame temperatures for various flame types on premix burners are given. It is not correct to leave the impression that a flame burns at a certain temperature; hence the word "about" was used in front of the word "temperature."

Depending on the proportions of oxidant (usually air or nitrous oxide) and fuel (usually hydrogen or acetylene) the temperature of a particular flame can be changed slightly. Fuel-rich flames are cooler than oxidant rich flames.

It is important to consider flame composition when chemical interference is a problem. In the case of elements that form stable oxides, e.g., calcium, silicon, and titanium, a reducing (fuel-rich) flame is best. The higher-temperature nitrous oxide–acetylene flame also minimizes this problem.

Techniques Involving Use of a Flame for Heat to Produce Atomization in a Sampling Vessel

Better detection limits, compared to conventional flame atomization, are claimed for silver, arsenic, bismuth, cadmium, mercury, lead, selenium, tellurium, thallium, and zinc in certain sample types using flame-heated sampling vessels. This approach was first introduced by Delves (11). The sample is inserted into the vessel and concentrated by drying. Atomization is accomplished by heating the vessel to a high temperature in a flame. The apparatus is inexpensive and is a compromise between flame and expensive electrothermal devices. It should also be noted that it is extremely difficult to obtain good reproducibility using this approach.

The Perkin-Elmer offerings in this area are typical. A Delves microsampling cup (Fig. 16) and a sampling boat are available. These metallic sample holders can contain up to 0.1 and 1 ml of sample, respectively. The sample is first dried using a heat lamp, hot plate, or flame, and then inserted into an air–acetylene flame on a three-slot burner. The holders, devices for supporting and aligning the system, must be treated carefully since precision placement of the vessel in the optical beam will mean the difference between success and failure.

The Delves microsampling cup, originally designed for lead in blood determinations, consists of a small nickel crucible, capacity 0.1 ml. A quartz absorption tube with an entrance hole in the bottom is heated in an air–acetylene flame. The cup is positioned just below the entrance hole of the hot tube. The atomized sample passes into the tube and

Fig. 16. Delves Micro Sampling cup as offered by Perkin-Elmer Corp.

because of an increased residence time of the atoms in the tube (optical path) an enhanced sensitivity is obtained.

The sampling boat is a tantalum vessel with 1 ml capacity. Sample placed in the boat is dried next to the air–acetylene flame and then atomized by *precision* placing of the boat in the flame just below the optical beam.

Tantalum is superior to nickel when acid samples are used. However, the useful lifetime of the tantalum boat under acid conditions, due to oxidation, is only 20 to 50 determinations. When acid is low, nickel cups are usable at least 40 times. As the vessels become older a decrease in sensitivity is noted. Replacement is recommended after sensitivity has decreased to 50% of its initial value.

Chemical and physical interference problems are more severe with these systems. This is a result of the cup and boat not achieving the temperature of the flame. Constant flame temperature is also important. Water should be nebulized throughout the analysis to minimize this problem.

Absorption signals appear in the form of relatively sharp peaks. Peak shape for a given element is influenced by the condition of the inside surface of the cup or boat, the position of the cup or boat in the flame, and the nature of the sample matrix. A recorder must be used with the Delves cup or sampling boat systems to facilitate diagnosis of interference problems.

Electrothermal Atomizers

There is no area of atomic absorption spectroscopy research that presently receives as much attention as electrothermal atomization. A book by Fuller (12) documents much of this work.

L'vov (13) introduced electrothermal atomizers into atomic absorption analysis. The term "electrothermal atomizer" will be used in this book to refer to electrically heated devices such as graphite furnaces and rods. The common designation, flameless or nonflame atomizer, will not be employed to avoid implications that other devices such as cold vapor absorption tubes are being discussed in this section.

Electrothermal atomizers are profoundly more difficult to utilize than flames; hence the latter should be used when applicable. Interference problems, particularly nonspecific background types, are very severe. To date, as is the case with flames, electrothermal atomization mechanisms are poorly understood.

Most early commercial atomic absorption equipment was less than optimal for electrothermal work. Electronic systems failed to follow the very fast transient signals obtained with electrothermal atomization. Baffling was essential in many optical systems to cut down on stray light. While rapid progress was made in solving these problems, the main drawback with the electrothermal equipment itself, lack of real control of temperature, was not adequately addressed in commercial equipment until 1977. To minimize interferences it is often important to use the highest-power ramp available during the atomization cycle. With control of temperature, this can now be done without danger of overshooting the desired final temperature. Without temperature control capability, reproducible and accurately definable heating cycles were impossible. As a result, most of the thermal programs that now appear in the literature possess a degree of uncertainty. This will explain many of the disagreements between researchers on thermal volatilization programs and interferences. The new equipment does not accurately measure the temperature in the atomization zone, and hence disagreements, but at a much reduced level, will persist.

In spite of these negative comments the advent of electrothermal atomic absorption represents a milestone in the development of analytical methods. Because of increased atom residence times and conversion of nearly all the analyte in the sample into atoms, detection limits are improved by up to three orders of magnitude compared to flames. With the commercial production of devices with capability for real control of temperature, there now exists a more solid base for research into electrothermal atomization. In addition, the analyst can expect these new devices to be more reliable and useful in routine analysis.

There are two types of electrothermal device generally accepted today, the graphite tube furnace (Fig. 17), and the carbon rod (Fig. 18).

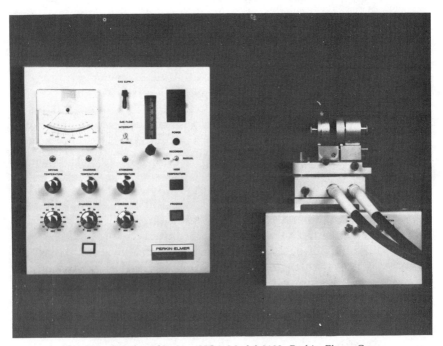

Fig. 17. Graphite furnace, HGA Model 2100, Perkin-Elmer Corp.

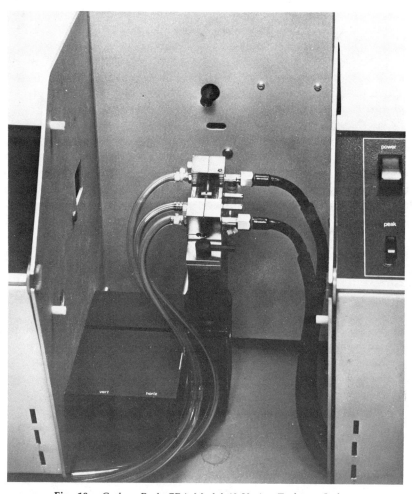

Fig. 18.   Carbon Rod, CRA Model 63 Varian Techtron Ltd.

Both involve samples atomized in a tube. In the case of the graphite furnace the purge gas is injected into and travels axially through the center of the tube. With the carbon rod, the purge gas is injected from below the rod/tube assembly and simply surrounds the assembly. In general, better limits of detection are possible with the furnace compared to the rod. Metallic strip atomizers were formerly available, commercially, but for many elements detection limits were poorer and interferences worse. A number of other electrothermal atomizers have been proposed, e.g., wire loops (14), carbon filaments (15), and a graphite T tube furnace (16). None of these have become generally accepted.

Electrothermal atomizers have a small sample capacity (1–100 μl). It is difficult to manually inject the microliter volumes reproducibly onto the atomizer. Autosamplers are now available for both the furnace and the rod and come highly recommended by the author.

Detection limits are up to three orders of magnitude better than with flames. Table II lists the elements often done by electrothermal atomization and their detection limits in solutions low in dissolved salts (i.e., interference-free matrix). Detection limits are, of course, matrix dependent. In complex high-salt-content solutions, the detection limits for most elements given in Table II are seriously degraded.

Principles of Operation

The electrothermal atomizer consists of the atomizer component, a power supply, electrodes that deliver power to the atomizer, and a casing or frame. The latter usually makes provision for cooling and inert gas sheathing of the atomizer.

The power supply normally provides three heating cycles. First, the sample can be dried at around 100°C. Charring, the volatilization of organic and low-boiling-point inorganic compounds, occurs in the second cycle. The highest temperature not resulting in loss of analyte is

Table II

Detection Limits (pg) for Analyte in Low-Salt[a]-Content Water, Perkin-Elmer 603 and HGA 2100 Using Conditions Recommended by Manufacturer

Element	Detection limit	Element	Detection limit	Element	Detection limit
Ag	0.5	Ir	1000	Se	200
As	25	Mn	0.8	Sn	200
Au	25	Mo	15	Te	100
Bi	30	Ni	30	Ti	500
Cd	0.5	Pb	5	V	200
Co	10	Pd	30	Zn	0.2
Cr	20	Pt	500		
Cu	5	Rh	50		
Fe	10	Sb	10		

[a] Interference—Free matrix.

generally chosen for this cycle. In the third cycle the device is heated to an appropriate atomization temperature of up to 3000°C depending on the analyte and the capability of the power supply. Each cycle can be programmed to last for the desired length of time. Modern commercial units commonly provide a continual and variable ramp heating program for each cycle.

The sample is usually placed in the atomizer as a liquid. A microliter pipet with plastic tip has been found to be best for sample introduction. Atomizers take up to 100 μl of sample, depending on the type and manufacturer's specifications. An inert gas, such as nitrogen, is used to purge or sheath the device. The sample is then dried, charred, and atomized.

Signals occur in the form of peaks and a chart recorder should be used to record the output. The shape of the peaks vary from element to element, with atomizing temperature, with atomizer condition, and as a result of matrix interactions with the analyte. A recorder tracing is invaluable in the diagnosis of interferences in electrothermal work.

The atomizer surface deteriorates with use. As a result, two serious problems occur: (1) the peak height for a given concentration diminishes; (2) the resistance of the atomizer changes. Atomizer surface deterioration may be minimized by coating the graphite surface with a pyrolytic graphite layer. The coating is applied during manufacture or, in the case of graphite furnaces, the coating can be generated and regenerated *in situ* using a gas mixture containing methane (16a). The pyrolytic layer prevents the sample from soaking into the graphite, giving more reproducible atomization for carbide-forming elements and prolonging the useful lifetime of the atomizer. For most elements better sensitivity is obtained with pyrolytic graphite.

With much existing equipment, a change in the resistance of the atomizer yields erroneous temperature readings. In these devices, temperature is not actually measured. The "temperature" meter is really a current meter. Most recent instrumentation has temperature control.

The Atomization Process

No generally agreed upon quantitative model of atomization for electrothermal devices exists, despite volumes of research on this subject. The answer seems to lie in a combination of thermodynamic and kinetic considerations. To date, it appears that atomization could result from either reduction of metal compounds by carbon or thermal dissociation of metal compounds, usually metal oxides.

Atomization by Cathodic Sputtering

Of the many other atomizers that have been proposed, the sputtering device proposed by Gough *et al.* (16b) seems to have greatest potential in practical atomic absorption analysis (Fig. 19). In this approach the sample is made the cathode. Atoms are produced in a low-temperature, low-pressure discharge. Inert-gas ions produced in the discharge accelerate toward the cathode and knock out atoms from the sample. The approach is best suited to analysis of conducting materials. Nonconducting samples can be incorporated in a conducting matrix such as copper or silver and the sputtering method utilized. However, this approach cannot yet be recommended for routine work. Atomization in low-pressure discharges is potentially very useful for the atomic absorption analysis of the elements nitrogen, phosphorus, and sulfur. With the exception of indirect methods, which the author does not recommend, and for the determination of phospnorus at high concentration, these elements cannot be analyzed by conventional atomic absorption.

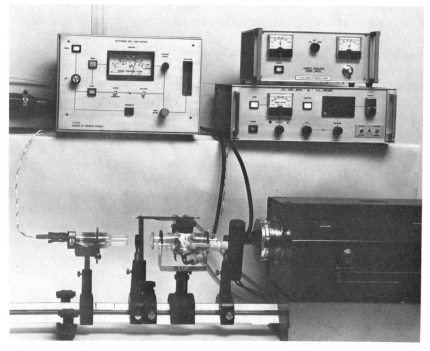

Fig. 19. Instrumentation for cathodic sputtering, CSIRO, Division of Chemical Physics, Melbourne, Australia.

Atomizers for Mercury and the Hydride-Forming Elements

High-temperature furnaces or flames are required to produce atoms of most elements in sample solutions. After appropriate chemical pre-treatment, mercury and the elements that form hydrides can be atomized at lower temperatures ($<1000°C$).

One of the major contributions in the analysis of geological and environmental samples by atomic absorption was the development of the cold-vapor absorption tube method for mercury by Hatch and Ott (17). This approach, based on a colorimetric procedure published by Kimura and Miller (18), allowed the determination of mercury in a wide variety of matrices at levels three orders of magnitude lower than atomic absorption methods available at that time. These latter methods were inadequate for the determination of the very low levels of mercury found in almost all geological and environmental samples.

Elemental mercury has an appreciable vapor pressure at room temperature. Thus absorption measurements can be made by sweeping mercury vapor into an unheated absorption tube placed in the optical beam of the atomic absorption unit. Tubes of a variety of sizes and compositions have been used. The author employs a plexiglass tube 10 cm long by 2 cm i.d. fitted with quartz end windows. Background correction should be used when analyzing mercury by this method. This negates the use of inexpensive, commercially available mercury analyzers, which do not have this capability.

The Marsh reaction for the generation of arsine has been known for 130 years. Another major contribution to atomic absorption analysis was the development by Holak (19) of an atomic absorption method for the analysis of arsenic based on the production of arsine. Arsine is atomized in an argon-entrained air–hydrogen flame or a low-temperature quartz furnace (850–950°C), the latter being generally preferred due to excessive absorption by flame gases of the useful arsenic resonance line at 1937 Å. Like mercury, with the advent of this method, levels of arsenic could be determined that were three orders of magnitude lower than were possible with procedures available at the time. The hydride approach has since been expanded to include the elements tin, germanium, bismuth, selenium, antimony, and tellurium. In contrast to mercury, methods have been developed using conventional electrothermal atomization, which can be used for analyzing covalent hydride elements in a number of sample types with detection limits comparable to the hydride method.

OPTICAL SYSTEM

Because of the simplicity of the spectrum emitted by most radiation line sources, the monochromator in atomic absorption spectroscopy need not be of the high resolution required for emission work. In atomic absorption the monochromator is placed after the atom reservoir to help diminish light flux on the phototube. The purpose of the monochromator is to isolate the resonance line and to decrease the emission intensity from the atomizer. Both prisms and gratings have been used in commercial instruments but the latter are of much greater popularity. When gratings are used, two are commonly employed, one for the ultraviolet and the other for the visible regions of the spectrum.

Atomic absorption spectrometers are either double or single beam. In double-beam instruments (Fig. 20) the beam of radiation from the source (S) is split by a chopper (C_C), one half going through the flame (B) and the other half bypassing it. The halves are combined using a beam-splitting mirror (C_M). Double-beam instruments lose a factor of 2 in signal as a result of this process. The main purpose of a double-beam optical system is to cancel instabilities in the source. With modern hollow cathodes, stability is generally very good. However, unless long warm-up periods are used, such is not the case for electrodeless discharge and vapor discharge lamp sources. When the latter are used a double-beam optical system is to be preferred.

Both mirrors and lenses are used in atomic absorption equipment. These devices must be capable of reflecting or transmitting light at wavelengths in the ultraviolet. The focal length of a lens depends on wavelength. Thus, a lens-optical system focused for ultraviolet wavelengths will not be in perfect focus for visible wavelengths. In practice, this aberration is not too serious. The above is not a problem with mirrors. The latter devices are easily damaged by chemical fumes, a problem that requires replacement at high cost. The surfaces of both mirrors and lenses must be kept clean to help avoid needless loss in beam intensity.

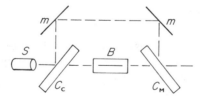

Fig. 20. Double-beam optical system.

The f number of the optical system is often quoted,

$$f = F/d$$

where F is the focal length and d the lens diameter. Thus, a high f number optical system has small diameter optics with long focal lengths. High f number optics are useful in applications involving the furnace because a narrow optical beam best fits through the furnace aperture.

The slits of the monochromator determine the amount of the spectrum that falls on the photomultiplier (band pass). It is generally advisable to keep the slits as narrow as possible to lower the flame emission reaching the photomultiplier and to reject unwanted lines from the hollow-cathode lamp. For example, in the determination of nickel at 2320 Å the slits should be adjusted to give a bandpass of less than 1 Å to minimize problems from nonabsorbing lines. The minimum slit width is determined by the amount to which the resonance line signal can be reduced before noise becomes too intolerable.

When purchasing equipment the question of whether to buy a double- or single-beam instrument arises. Double-beam systems correct for fluctuations in lamp output. This allows the worker to turn on the lamp and begin the analysis without waiting for the lamp output to stabilize. However, because of the more complex optics needed for this system, a serious penalty of reduced signal strength must be paid. The author believes that because of the high stability of modern lamps and the availability of warm-up turrets, standard with most single-beam instruments, the latter system may be purchased without fear of problems.

Figure 21 shows a grating monochromator using a Czerny–Turner mounting. Light comes in through the entrance slit (S_E), falls on the grating by reflection from a mirror (m), and is reflected out through the exit slit (S_x) onto a photomultiplier (PM). The grating is a flat plate coated with reflective aluminum. The surface is scored with parallel grooves at a density of 450–2900 grooves/mm. Light striking the grating is dispersed at an angle characteristic of the wavelength of the light. The greater the line density on a grating, the higher the angle of dispersion.

In contrast to a prism, the grating disperses light in a linear fashion over the wavelength range. Unfortunately, gratings produce second- and third-order lines of a given wavelength. These become troublesome for the analysis of elements in the high-wavelength range. To overcome this problem, a grating instrument is often equipped with a filter, which is used for wavelengths above 6000 Å. The determination of

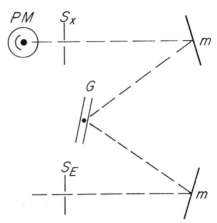

Fig. 21. Grating monochromator, photomultiplier system.

potassium using the 7665 Å line is a commonly encountered example of the need for this filter.

The f number of the monochromator is determined by the grating size and focal length. This number determines the solid angle of incident radiation that can be focused through the slit onto the grating.

ELECTRONIC SIGNAL PROCESSING

A photomultiplier tube sensitive to radiation over the wavelength range 1900–8000 Å is commonly used. Signals generated by this device are very small (in the nanoampere range). The photomultiplier is a current source. The intensity of the current produced is proportional to signal strength. The I_0 and I beams have been recombined and are separated in time (Fig. 22) by anywhere from 1 to 10 msec, depending on chopper rotational speed.

Current from the photomultiplier is converted to a voltage using a high resistance in the feedback loop of a high-input-impedance operational amplifier. This signal is amplified to a suitable level by the main amplifier. A reference waveform, often produced by a small lamp shining on the chopper, is used to switch the reference and absorbed beams into the respective demodulators. These devices produce signals I_0 and I proportional to pulse heights of the respective signals. Log amplifiers are used to process the signal. The signal is then fed into the differential inputs of an operational amplifier, which performs the required subtraction as follows:

$$A = \log(I_0/I) = \log I_0 - \log I$$

and a readout in absorbance is thus obtained.

Most equipment also has a built-in curve corrector that can be employed to straighten nonlinear calibration graphs at high levels of concentration. It is important that these curve correctors have a threshold control allowing the worker to adjust the curve only in that region in which it is nonlinear. It is unwise to run samples whose absorbances are in the upper region of the curved segments of the calibration graph. This is because the absorbance per unit concentration has fallen to such a small level that the uncertainty goes up very sharply. Curve correction is particularly well handled with microprocessor-controlled instrumentation.

Modern atomic absorption equipment provides a wide choice of signal integration time and scale expansion interval. In the case of signal integration time, a range from about 0.2 to 90 sec is adequate for most applications. The shortest integration time is limited by the time constant of the electronics and may sometimes be too long to record, without distortion, transient signals obtained from an electrothermal atomizer (caution: older equipment may be particularly deficient in this regard). Most often, however, longer integration times are desirable to optimize the S/N ratio.

Older instruments have meter readout and often provision for a recorder. Output from older instruments may not be linear with concentration, e.g., percentage absorption, and must be manually converted to absorbance. New equipment usually employs digital electronics allowing direct readout in concentration.

Microprocessors have recently been introduced into atomic absorption equipment. To date, these have been used mainly for setting up integration time and scale expansion and for instrument calibration and

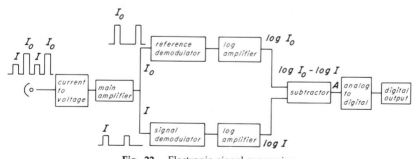

Fig. 22. Electronic signal processing.

curve correction. Instruments with complete microprocessor control are currently being manufactured. This development will have a large impact on ease of instrument operation and data handling. It is certain that microprocessors will revolutionize atomic absorption equipment design in the next few years.

CHOICE OF THE ANALYTICAL LINE

Resonance transitions are those which occur when a valence electron is excited from the ground state. Most lines used in atomic absorption spectroscopy are of this type.

The resonance transition requiring the least energy, i.e., that transition from the ground state to the lowest-energy excited state, usually gives the best sensitivity and is commonly chosen for atomic absorption work. In other cases where sensitivity is not so important, the most sensitive line is still frequently used. This is because the sample can easily be diluted to give concentrations within the working range of this line. Sensitive lines, being more intense and stable, are usually chosen unless a secondary line of good characteristics is available.

For some elements good secondary lines have sensitivities that are conveniently a few times less, e.g., the 3415 Å nickel line is four times less sensitive than the 2320 Å, the most-sensitive nickel line. Others have great sensitivity differences between lines, e.g., the most sensitive zinc line, 2139 Å, is 4700 times more sensitive than the 3076 Å line, which is the only other suitable zinc line.

Lines generally useful for atomic absorption must be between 1900 and 8500 Å (9000 Å with a suitable photomultiplier). This is because below 1900 Å, absorption of the source radiation by air is almost complete. Unfortunately, the resonance lines for important elements such as sulfur, phosphorus, and nitrogen are below this wavelength and can only be used with special vacuum equipment.

In the wavelength region 1900–2000 Å absorption of source radiation by air and flames is high. The best flame for arsenic determination, at 1937 Å, is air–hydrogen. But even with this "best" flame, detection limits are poor.

Table III lists the most sensitive lines for each element plus a selection of other lines and their approximate relative sensitivity. The worker is cautioned that hollow-cathode lamps and other sources used may not operate stably at all these wavelengths.

Table III

Selected Lines Suitable for Use

Metal	Wavelength (Å)	Relative sensitivity[a]	Metal	(Å)	Wavelength Relative sensitivity[a]
Ag	3281	1	Ni	2320	1
	3383	2		3415	4
As	1937	1		3515	10
	1972	2		2944	50
Au	2428	1	Os	2909	1
	2676	2		2807	5
	3123	1000		4261	30
Bi	2231	1	Pb	2833	1
	3068	4		2614	10
	2277	15		3684	25
Cd	2288	1	Pd	2476	1
	3261	400		3405	3
Co	2407	1	Pt	2659	1
	2521	2		3065	2
	3044	10		2719	8
	3018	100	Rh	3435	1
Cr	3579	1		3701	10
	3605	2	Ru	3499	1
	4290	5		3799	2
Cu	3247	1		3926	10
	3274	2	Sb	2176	1
	2226	15		2312	2
	2244	150	Se	1960	1
Fe	2483	1		2040	3
	2527	5		2063	10
	3737	10	Sn	2246	1
	3466	100		2863	2
Ir	2640	1		2661	30
	2544	2	Te	2143	1
	3514	10		2259	15
Mn	2795	1	V	3184	1
	2801	2		3902	7
	4031	10	Zn	2139	1
Mo	3133	1		3076	4700
	3194	2			
	3112	20			

[a] 1 is the highest sensitivity.

INTERFERENCES

Nonanalytical spectroscopists are often unfamiliar with interference problems in atomic absorption spectroscopy. Many analysts, when they begin using this technique, are of the opinion that interference patterns are simple and predictable. While this may be true for a few elements such as copper, the vast majority have interference problems that are complex and not always predictable. This has very serious consequences, particularly in work with electrothermal atomizers.

Atomic absorption spectroscopy has only been available as an analytical tool since 1955. During the first eight years few practical applications were published. Only since about 1964 has there been a proliferation of analytical applications investigated. For this reason it is important to stress that quantitative information about interferences in many cases may still be lacking. Even more disturbing is the well-known fact that many authors disagree strongly about specific interference patterns.

The reason for the disagreement is due largely to the use of different equipment and experimental conditions. It has become abundantly clear that the optical beam height in the flame, flame composition, flame configuration, etc., are factors that strongly affect the interference pattern. In a test in the author's laboratory (20), even with operating conditions adjusted to be as closely similar as possible, different interference patterns were obtained for molybdenum using two different brands of atomic absorption equipment.

Some trace metals, e.g., copper and zinc, do not have serious problems associated with their determination even over a fairly wide range of sample types. However, when trace levels of any element are done in the presence of very high concentrations of a matrix constituent it is possible that a chemical interference may occur. In such cases, it is often best to use matrix standards or use the method of standard additions.

In high-salt-content solutions, nonspecific background interference can readily occur for any trace metal. This problem is particularly severe with electrothermal atomization. Under these conditions background correction must be employed.

Interference problems in electrothermal atomizers are very much more poorly understood than in flames. Difficulties in the study of interferences in these devices are compounded by the lack of a quantitative description of the atomization process. Considering that interferences are the limiting factor in use of electrothermal atomizers, the situation is critical to say the least.

Types of Interference

Atomic Spectral Interferences

This type of problem occurs when the radiation being measured is attenuated by substances other than the atoms of interest. Atomic spectral interference occurs when the absorption profile of another element overlaps that of the analyte within the spectral linewidth of the emission line of the source. Because of the narrow width of atomic emission lines, atomic spectral interference causes very few problems.

At first glance the reader may wonder why radiation emitted from the atomizer at the analyte wavelength would not be added to the resonance emission wavelength from the hollow-cathode lamp to somewhat mask the absorption effect. This does not occur in modern instruments since the source radiation is modulated and the amplifier used to detect the signal from the photomultiplier is tuned to receive only that modulated frequency. In this way emission from the flame or electrothermal device is not detected by the amplifier. The latter, however, may cause noise problems.

A number of atomic spectral interferences have been reported. These have been compiled by Norris and West (21) and are given in Table IV (21a, 22). Fortunately, atomic spectral interferences are rarely encountered in the types of analyses done in most laboratories.

Nonspecific Interferences

Nonspecific interference occurs due to molecular absorption or light scattering in the atomizer. These effects, seldom encountered in flames except in work near the detection limit, are a very serious problem in most work with electrothermal atomizers. In the case of the latter, nonspecific absorption often minimizes the detection limit advantage compared to flames.

Molecular absorption occurs when a molecular species in the atomizer has an absorption profile that overlaps that of the element of interest. This problem is most serious in the wavelength region below 2500 Å. Molecular absorption profiles are relatively broad compared to atomic absorption profiles and usually have narrow rotational lines superimposed on this continuum. If one of the latter coincides with the atomic absorption line, background correction cannot be done.

Light scattering occurs when solid particles cause a deflection of some of the source radiation away from the axis of the monochromator–detector system. This results in a decrease in beam intensity, a

Table IV

Cases of Spectral Overlap Reported for Atomic Absorption

Source[a]	Emission wavelength (Å)	Analyte	Absorption wavelength (Å)	Separation (Å)	Sensitivity (ppm)[c]
Iron	3247.28	Copper	3247.54	00.26	0.80 (a)
Iron	3274.45	Copper	3273.96	00.49	1.10 (a)
Iron	2852.13	Magnesium	2852.13	<00.01	10.00 (a)
Iron	3524.24	Nickel	3624.54	00.30	0.10 (a)
Iron	2794.70	Manganese	2794.82	00.12	0.04 (a)
Copper	3247.54[b]	Europium	3247.53	00.01	75.00 (b)
Iron	2719.03[b]	Platinum	2119.04	00.01	40.00 (b)
Silicon	2506.90[b]	Vanadium	2506.90	<00.01	65.00 (b)
Aluminum	3082.15[b]	Vanadium	3082.11	00.04	800.00 (b)
Gallium	4032.98[b]	Manganese	4033.07	00.09	15.00 (b)
Manganese	4033.07[b]	Gallium	4032.98	00.09	35.00 (b)
Mercury	2536.52[b]	Cobalt	2536.49	00.03	100.00 (b)
Germanium	4226.57	Calcium	4226.73	00.16	6.00
Neon	3593.52	Chromium	3593.49	00.03	14.00 (b)
Neon[d]	3593.52	Chromium	3593.49	00.03	0.10
Antimony	2170.23	Lead	2169.99	00.24	250.00 (b)
Iodine[d]	2061.63	Bismuth	2061.70	00.07	10.00
Arsenic[d]	2288.12	Cadmium	2288.02	00.10	45.00
Zinc	2138.56[b]	Iron	2138.59	00.03	200.00 (b)
Iron[e]	2874.17[b]	Gallium	2874.24	00.07	250.00
Iron[e]	4607.65	Strontium	4607.33	00.32	20.00
Iron[e]	3382.41	Silver	3382.89	00.48	150.00
Iron[e]	3961.14	Aluminum	3961.53	00.39	50.00
Mercury[f]	3593.48	Chromium	3593.49	00.01	250.00
Mercury[f]	2852.42	Magnesium	2852.13	00.29	200.00
Lead	2411.73	Cobalt	2411.62	00.11	15.00
Lead	2476.38	Palladium	2476.43	00.05	3.50
Antimony	2169.19	Copper	2178.94	00.25	100.00
Antimony	3232.52	Lithium	3232.61	00.09	200.00
Antimony	2311.47[b]	Nickel	2310.95	00.52	35.00

[a] Sources were hollow-cathode lamps, except as noted.

[b] Resonance lines.

[c] Figures marked (a) are reported detection limits, those marked (b) are sensitivities calculated from the data given in the papers, and the remainder are reported sensitivities. Sensitivity is defined here as the concentration (ppm) required to produce an absorbance of 0.0044 at the specified wavelength and under the conditions specified in the original papers.

[d] Electrodeless discharge lamp.

[e] Suggested by Frank et al. (21a).

[f] Reported in atomic fluorescence by Omenetto and Rossi (22).

pseudoabsorption effect. Light-scattering problems are most prevalent with samples containing high levels of refractory elements.

Like molecular absorption, light-scattering interference is greatest at wavelengths below 2500 Å. This problem is, however, frequently encountered at wavelengths well into the visible region of the spectrum.

Correction for nonspecific absorption is described on p. 44. It is important to test the need for background correction in all trace metal analyses when using electrothermal atomizers and with flames when working near the detection limit. In the case of the former background, correction is almost always necessary.

Chemical Interferences

During the early years of atomic absorption many workers failed to recognize the serious problems caused by chemical interference. These problems, well documented in emission work, were largely unknown to the flood of nonspectroscopists who were attracted to atomic absorption.

Chemical interference occurs when the analyte is contained in a chemical compound that is not broken down by the flame or furnace. This results in a lower concentration of "free" analyte atoms than would occur in the absence of the interferent. Atomic absorption, of course, can only occur by free atoms.

A good example of chemical interference is the reaction of calcium and phosphate ions during evaporation of liquid droplets in the flame to form calcium phosphate. This compound, with heat, is converted to calcium pyrophosphate, which is relatively stable at the temperature of an air–acetylene flame. The chemical reaction reduces the free calcium atom population in the flame compared to that obtained for similar calcium solutions without phosphate.

In work with flames, chemical interferences can be minimized either by using a higher-temperature flame and/or through addition of a releasing agent. In some instances releasing agents preferentially combine with the interfering ion, thus releasing the analyte atom. In other cases, combustible releasing agents such as EDTA combine preferentially with the analyte atom and then burn in the flame-releasing analyte atoms.

In the calcium example, the phosphate interference can be alleviated by either adding a releasing agent or by using the higher-temperature nitrous oxide–acetylene flame. Lanthanum or strontium can be added in large amount (~1%) to preferentially bind phosphate, thus releasing calcium for atomic absorption, or EDTA can be added to preferentially react with calcium. Once released calcium does not recombine with the phosphate in the interconal zone of the flame.

It is important to note that although higher-temperature flames do successfully overcome some chemical interferences they do not alleviate all such problems. The author has found that the depressive effect of zirconium on yttrium absorbance could not be fully overcome in the nitrous oxide–acetylene flame even in the presence of a releasing agent.

The method of standard additions is useful for overcoming chemical interference particularly in electrothermal work. This technique is, however, time consuming compared to the above approach.

Ionization Interferences

Atoms of elements possessing very low ionization potentials can become ionized at flame and furnace temperatures. This interference, also known as vapor phase interference, reduces the free-atom population. A calibration curve, as shown in Fig. 8, line C, is obtained when ionization interference is experienced. Elements that commonly present a problem are cesium, rubidium, potassium, and sodium. At nitrous oxide–acetylene flame temperatures calcium, rare earths, strontium, and barium are also ionized to an appreciable extent.

The magnitude of ionization in a flame depends on the concomitants present and their concentrations. If an analyst is determining potassium in a sample high in sodium, the ionization of potassium is greatly suppressed by the presence of sodium. Hence the fraction of potassium ions produced in a 1 ppm sample solution of this type will be much less than in a 1 ppm standard not containing sodium. To prevent this inequality of ionized species between samples and standards, a large excess of an alkali metal (1000 ppm) should be present in all solutions.

Physical Interferences

When the physical characteristics, e.g., viscosity and surface tension, of sample and standard solutions are different, an interference in the condensed phase is experienced. For example, in work with burners, solutions of different viscosities will pass up the capillary into the nebulizer at different rates. Different free-atom populations will result, in the flame, for the same analyte concentration in the solutions.

When micropipets are used to introduce sample into an electrothermal atomizer, the bulk composition of the solution will affect the volume dispensed. The distribution of sample in the electrothermal atomizer is also dependent on solution composition. A physical interference can result if sample and standard solutions vary greatly in bulk composition.

Special Considerations for Electrothermal Atomization

Volatilization Interference

Interferences in electrothermal work can be very complex. This is particularly true with heavy matrix samples. For overcoming physical and chemical interferences, the method of standard additions can often be employed.

In general, high-volatility elements such as lead, cadmium, and zinc can be atomized at relatively low temperatures, giving sharp atomization peaks. Low-volatility elements, e.g., nickel, vanadium, and molybdenum, are released into the optical beam at temperatures in the upper range of the electrothermal atomizer, yielding broader atomization peaks. Coexisting cations and anions can affect the rate of volatilization of an element and thus alter the shape of the atomization peak.

Any phenomenon that alters the rate of volatilization of analyte from a sample compared to a standard is an interference. Interferences of this type are very poorly understood. To minimize the problem it may be possible to change the composition of the matrix either by reagent addition or by selective volatilization of matrix components prior to atomization. Signal integration or standard addition are also employed. For the latter approach to be valid, the added analyte must assume the same chemical form in the atomizer as the sample analyte. Standard addition also requires that the calibration graph be linear over the concentration range employed. Calibration graphs can have very short linear regions in electrothermal work. Of course, using standard addition is time consuming, but for the most accurate work it may be essential.

There is a danger that, during the heating cycle, analyte can be lost prior to the atomization step. This is a particularly serious potential source of error when chloride is present. It is advisable, therefore, to convert all chlorides to a less volatile form, e.g., nitrate or sulfate, prior to the ashing step. A number of elements such as arsenic, selenium, zinc, cadmium, and lead are potentially volatile at relatively low temperatures. It may be possible in these cases to vaporize the analyte prior to the matrix. When this approach is not valid, recent research has shown that an element such as nickel (23) (in the case of selenium or tellurium) can be added to stabilize the element in the atomizer during the ashing step. Other transition elements such as iron have the same effect; hence in many geological samples, elements such as arsenic, selenium, and tellurium may be stabilized by the existing matrix.

To complicate matters, analyte can undergo complex reactions after volatilization, i.e., in the gas phase. This matrix-dependent, poorly understood chemical interference is best handled by standard addition.

Nonspecific Absorption Interferences

Without doubt, the most important interference using electrothermal atomizers results from signals due to light scattering or molecular absorption. This problem, most severe at ultraviolet wavelengths, also must be considered in the visible region of the spectrum. The effect can be so acute that the background correction capability of the instrument is exceeded. Under this circumstance, much of the older equipment will still generate values without any indication of error. The consequences of such a phenomenon are obvious. If there are no warning lights indicating background correction overload, but the equipment has an energy meter, the observation that the energy meter needle drops drastically toward zero can be evidence of the problem. Most commercial atomic absorption units cannot background-correct when the interference produces a signal greater than an absorbance of 1. Recent equipment embodies a warning light to indicate when the background correction capability has been exceeded.

Matrix modification and/or a background correction method are used to overcome nonspecific interferences. Compounds such as sodium and potassium chlorides and organic materials come off at relatively low temperature. Their generation can interfere with the determination of low-boiling-point elements (cadmium, lead, zinc, etc.). For high-boiling-point elements compounds such as the alkali metal salts can be reduced to tolerable levels by selective volatilization. The presence or formation of refractory oxides can interfere severely with the determination of high-boiling-point metals. Selective volatilization is not possible in this case. It must be emphasized that the method of standard additions is not applicable to overcoming nonspecific interferences.

Background Correction

One of the most challenging problems for the analyst involved in trace metal, atomic absorption spectroscopy is to make a valid correction for nonspecific absorption. Even with the best equipment available, at the time of writing this volume, background correction requires the attention of a skilled operator.

As already stressed, background problems are most severe with electrothermal atomizers. Selective volatilization of the offending matrix constituent from the analyte or vice versa can sometimes be used to minimize nonspecific absorption. This approach is only useful when the two substances in question have large differences in boiling points.

Figure 23 depicts background absorption in the region of a resonance line with wavelength λ_0. The magnitude of the absorption obtained will be the sum of the atomic absorption and the nonspecific absorption at the resonance wavelength. A correction to the absorption must be made by subtracting the background. The two approaches most commonly used to overcome this problem are the nonabsorbing-line and the continuum methods.

Nonabsorbing-Line Method

Emission from hollow-cathode lamps contains not only resonance lines of the element of interest, but frequently nonabsorbing lines of this element and lines resulting from excitation of the filler gas. In choosing a nonabsorbing line suitable for background correction, two important questions must be asked: (1) Is the line close to the resonance line (but at least two bandpasses away to avoid overlap)? (2) Does it give any absorption for up to 100× the analyte concentration being used? The nonabsorbing line may be in the lamp containing the analyte element or in a separate lamp.

Regarding the closeness of the nonabsorbing line to the resonance line, the following should be noted. Background effects may not be constant with wavelength. In Fig. 23 note that the background is quite uneven. For correction in the region AB a nonabsorbing line must be close to the resonance line. In the region CD, a nonabsorbing line could be some distance away without appreciable error occurring. When working in the region AB it is advisable to use two nonabsorbing lines, one on either side of the resonance line if such can be found. A listing of good nonabsorbing lines for some elements is given in Table V. The worker can also find the lines that may be required for his own unique case, as already indicated.

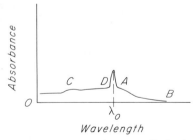

Fig. 23. Schematic representation of nonspecific background absorption.

Table V

Nonabsorbing Lines

Element	Analytical line (Å)	Background line (Å)
V	3184	3196
Cr	3579	(366x)[a]
Cd	2288	2264
Pb	2833	2820
Ni	2320	2316
Zn	2139	(212x)[a]
Cu	3247	2961
Mn	2795	2817
Ag	3281	3324

[a] Approximate value (accurate value not known).

To correct for background using the nonabsorbing-line method, the instrument characteristics must be considered. In dual-channel instruments it is possible to set one channel on the absorbing line and one on the nonabsorbing line. The background correction is then done by subtracting the background signal, either manually or automatically.

If a dual-channel instrument is not used, the whole set of samples must be run at the absorbing wavelength and then again at the nonabsorbing line. Subsequently a manual subtraction is done. Should flame conditions change during the run, a serious error can occur.

Continuum Method

Most manufacturers can supply a continuum emission source, usually a deuterium arc lamp. A good signal is obtained over the region 1900–2800 Å, but the arc can normally be employed at wavelengths up to about 3300 Å. It may also be possible to obtain a quartz iodide lamp for the visible wavelength range.

In the continuum method the signals obtained with a continuum and the line source are monitored separately. Their difference can give a background-corrected value. If narrow slit widths are used, significant absorption of the continuum radiation at the analyte wavelength may occur. This effect leads to a loss of sensitivity. Instruments that give dual beam background correction or the equivalent are to be preferred because of drift problems inherent to continuum sources.

To use background correctors, follow the manufacturer's instructions carefully. It is crucial to have the two beams optically coincident, filling the same fraction of the optical aperture and of roughly equal signal intensity. A good method for ensuring that the background correction system is operating properly is as follows: insert a wire screen having an absorbance equivalent of about 0.4 into the optical beams. An absorbance reading of zero should be obtained. If this is not the case, readjust the beam alignments.

Hydrogen lamps, used as continuum sources with some older equipment, are difficult to employ for background correction because of their relatively low output intensity over the spectral band pass compared to a hollow-cathode line source. Workers must also be cautious because hydrogen lamps sometimes generate nickel line radiation and hence will cause reduced sensitivity if used in a nickel analysis.

At narrow slit widths and when the arc is nearing the end of its useful lifetime, the continuum signal is drastically reduced. It may be necessary under this condition to reduce the hollow-cathode lamp current to allow the output from both sources to be balanced.

It may be found that the concentration readings obtained are either slightly higher or lower than expected. This problem commonly results from a failure to have the two signal beams optically coincident and filling the same fraction of the optical aperture. In some equipment the two beams may appear to be optically coincident, while still not filling the same fraction of the slit aperture.

As noted earlier, if molecular absorption is experienced and if one of the narrow rotational lines is coincident with the atomic absorption line, background correction by this method will be in error. This will result in over- or undercorrection.

Most automatic background correctors will compensate for signals up to an absorbance of 1. When this value is exceeded the sample must be treated, e.g., by dilution or extraction, to reduce the magnitude of the effect.

If further reading is desired on problems of background correction using a continuum source, the paper by Hendrikx-Jongerius and de Galan (24) is recommended.

Zeeman Background Correction

The Zeeman effect can be employed for background correction. In this method a magnetic field, applied to the source or the atomizer, is used to split the resonance line into its Zeeman components (π and $\pm\sigma$) as shown in Fig. 24 (24a). Using polarizers, background is monitored on the wings ($\pm\sigma$) and the analyte signal plus background with the central component. This method, available on one commercial furnace atomic absorption unit, has failed to obtain general popularity among instrument manufacturers, for the following reasons: splitting may be accom-

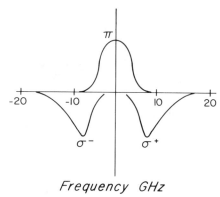

Frequency GHz

Fig. 24. Zeeman components of a resonance line (24a).

panied by less than 100% depth of modulation (loss of sensitivity); the elements require different magnetic fields to achieve maximum depth of modulation; the magnet has to date only successfully been applied around the atomizer and this is cumbersome and undesirable.

Caution

It has been implied throughout this section that the above methods give a true background-corrected value. Such will not occur with any of the above methods when there is direct spectral overlap between the fine structure of a molecular absorption band and the analyte line (24).

CALIBRATION PROCEDURES

There are two commonly employed approaches to calibration in atomic absorption spectroscopy. These are the methods of direct comparison and standard additions.

Direct Comparison

This is the simplest approach, and when applicable can be used with many instruments to give a direct readout of the concentration of an element in an unknown sample.

To obtain good precision, e.g., 1–2% coefficient of variation, the absorbance levels measured should be between about 0.1 and 0.6 units. An integration interval of about 2 sec should be used. Standard and sample solutions should be similar in bulk matrix constituents, particularly acid and total salt content. Interference suppressants must be in all solutions when required.

A number of standards (usually three to five) in increasing concentrations, as well as a blank, are prepared to cover the concentration range. These solutions are run in absorbance to check for linearity of the calibration curve. If curvature is present this fact is noted and with most modern instruments correction can be made to allow direct readout.

With older equipment, it may not be possible to have the instrument read out directly in concentration. In these cases the output is obtained in absorbance, percentage absorption, etc. Since absorbance is the function bearing a linear relation to concentration, all readout must be converted to absorbance units. It is important to note that, before doing so, readout from some instruments must be divided by the scale expansion factor.

When absorbances have been obtained, these values (A_1, A_2, A_3 in Fig. 25) (25) can then be plotted against concentration (C_1, C_2, C_3) for standards and a "best fit" line drawn. The portion of the analytical line BD gives best results, as shown by the dashed line. In many cases the line thus produced will be straight. When curvature occurs, usually at higher concentrations, a French curve or other manual curve-fitting device can be used to draw the line. Much poorer results are obtained in the curved region of the calibration graph, and it is advisable, where possible, not to use this portion of the graph. Likewise at low values of absorbance, where the change of precision with concentration is high, bad precision and hence poorer results are obtained. If the calibration graphs goes through zero absorbance and concentration, as in Fig. 25, then all is well. However, should the line intercept on the absorbance

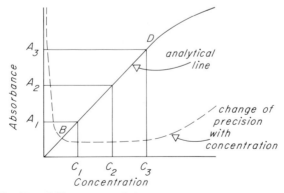

Fig. 25. Calibration graph for method of direct comparison.

axis after subtraction of the blank, an error due to nonspecific absorption may be present.

Most new equipment, including many of the less expensive models, has the capability for direct readout in concentration. It is very important to precheck calibration linearity. Any nonlinearity is readily detectable and strategy for curve correction can be devised. At least three standards should be used. The actual technique employed for direct readout and curve correction depends on the particular equipment. Instruments with microprocessors can curve-correct automatically. The reader must refer to the instrument manual for details.

It is not sufficient to read the standards only once during a run. For best accuracy, the author runs one of the standards at least three times over a set of 25 samples. This will allow a statistical analysis to be done. Means of the standard readings are used and linear regression analysis is possible.

In both the direct readout and plotting methods, the importance of using blanks cannot be overemphasized. In work at low concentration, the level of analyte in reagents is often an important consideration.

Output can be fed into desk top or large-scale computers. This can be done through manual methods or with proper interfacing devices, automatically. Again, the manufacturer's individual recommendations must be followed. Sometimes interfacing devices are available from the manufacturer, but it is still common to require custom work to provide this facility.

When an instrument is equipped with a microprocessor the calibration procedure is simplified. Experimental parameters are entered by keyboard. Any integration time or scale expansion factors between fixed upper and lower limits can be chosen. Values for the standards are entered and the standards subsequently nebulized. Linear regression and curve-fitting programs are carried out automatically. Samples are then nebulized and direct readout in concentration is obtained. Some recent instrumentation does other statistical calculations on the data as well. Better precision and accuracy are obtained in the curved region compared to the above manual methods. However, for best precision and accuracy, the linear portion should still be used.

Method of Additions

When samples contain a low concentration of analyte in large amounts of varying matrix constituents, it is often difficult to prepare useful standard solutions. In this case it may be possible to add small amounts of conventional standard solutions, in increasing amounts, to aliquots of each sample. A calibration graph such as that shown in Fig. 26 can then be constructed. The method of standard additions is very often used in work with electrothermal atomizers. For this method to work, the added analyte must behave during the atomization process in an identical way to the analyte that was originally present in the sample. This may not be the case, particularly using electrothermal techniques.

The calibration curve must be linear within the concentration range used. Such a condition commonly exists with flame techniques. However, it is well known that working curves for electrothermal methods, particularly in heavy matrix solutions, are often characterized by relatively short linear regions. The standard additions approach in this case can lead to serious errors.

The usual method of standard additions involves using three aliquots of the same sample. The first aliquot is diluted to volume while the second and third are spiked with increasing amounts of a standard and then diluted to a similar volume. It is best to spike samples such that the second solution has a concentration of approximately twice that of the unknown and the third three times the unknown. The absorbances are obtained, A_1, A_2, A_3, respectively, and a graph constructed as shown in Fig. 26. The line is extrapolated to intersect the x axis. While it may be possible with some equipment to obtain a direct readout in concentration, the procedure is time consuming and subject to limitations of the

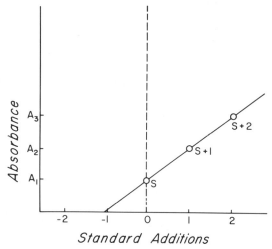

Fig. 26. Calibration graph for method of standard additions.

instrument in generating negative numbers. If nonspecific absorption is a problem, background correction must be employed.

The standard additions method cannot be used to overcome all interferences. Most particularly, nonspecific background is not corrected by this approach.

High-Precision Determinations

In general, despite claims to the contrary, atomic absorption cannot give routine results on the same sample taken through the complete procedure several times with a precision better than 0.5% coefficient of variation. More realistically, a 1% coefficient of variation is obtained, even with the most careful work. Such repeatability may not be acceptable in some applications. When higher precision is justifiable and required, it may be possible to obtain better values.

The most commonly employed approach is to prepare two standards that closely bracket the sample, i.e., ±5%. The lower standard is set to a suitably low value and then the scale is expanded such that the higher standard reads full scale. The sample is read. The values for standards are plotted on a graph and a straight line is drawn. The concentration of the sample is read from the graph.

ACCURACY AND PRECISION

The accuracy of a determination is a measure of how close the value obtained is to the "true" value. Unfortunately, the true value of a chemical constituent of a substance cannot be known absolutely; hence the assessment of accuracy is to a large extent subjective.

It is possible to develop standard reference materials that can be used to give an approximation of the accuracy of a determination. This requires the acquiring of large samples of the desired material, homogenizing the sample, and then undertaking a standardization program involving as many as possible different, but suitable analytical techniques. The data obtained are then processed statistically and a set of "accepted" values generated. Accuracy evaluated with standardized samples is termed "relative accuracy."

Precision of determinations is a measure of the degree of scatter obtained on replication of the analysis of a sample. Commonly, analysts obtain good precision with a method but have no data to allow an assessment of accuracy. Frequently good precision is mistakenly taken to indicate good accuracy. This problem is particularly evident among workers using atomic absorption spectroscopy. This statement is defended and the magnitude of the problem is emphasized in the following section.

A Critical Assessment of the Validity of Metal Data (26)

A review of the heavy-metal analytical literature shows that frequently there is no way to judge the validity of the numbers quoted. Reasons include poor documentation and control of sampling, sample preparation, and the subsequent analysis. In many instances, sampling may account for the greatest source of error. Indeed, sampling procedures have seldom received the same critical study as have methods of analysis; however, a discussion of sampling is beyond the scope of this book. Even assuming a good sample, it can be shown that heavy-metal chemical data on a particular parameter commonly vary from laboratory to laboratory by unacceptable amounts.

In the past five years there have been a growing number of interlaboratory comparison studies in heavy-metal analysis. It is crucial that samples prepared for this purpose be homogeneous and stable. In the examples quoted below the author feels that these requirements have been met as well as possible.

Mercury analyses have been done routinely for environmental and clinical purposes for much of this century. The numbers of analyses have increased dramatically since the Minamata mercury problems in Japan. Data accumulated during the mid-1950s and 1960s as a result of the Minamata incident demonstrated a lack of quantitative agreement between laboratories. In particular, large discrepancies were noted between different analytical techniques. One might expect that such a revelation would result in a rapid improvement in the quality of mercury data.

However, Uthe *et al.* (27) conducted an interlaboratory comparison of the analyses of mercury in fish tissue involving 29 laboratories. Table VI gives a summary of the data obtained on three fish samples by flameless atomic absorption spectroscopy (FAA) and neutron activation analysis (NA).

As can be seen, there is good agreement between the mean of FAA and NA. However, the range of results in each case indicates that individual laboratories disagree by as much as a factor of 7 (fish B).

Lauwerys *et al.* (28) recorded results of an interlaboratory comparison study involving 66 European laboratories. Mercury was analyzed in

Table VI

Mercury in Fish (μg/g)

		A	B	C
Mean	FAA	1.36 ± 0.23	0.10 ± 0.06	4.26 ± 0.77
Range	FAA	0.93 to 1.80	0.03 to 0.21	2.8 to 5.4
Mean	NA	1.37 ± 0.26	0.11 ± 0.06	4.06 ± 0.67
Range	NA	0.95 to 1.77	0.04 to 0.19	2.8 to 4.6

blood, urine, and aqueous solutions. Great care was exercised to ensure homogeneity of samples and in preparing and storing samples in an acceptable manner. Samples sent to laboratories were four identical portions of human urine (A), three identical samples of human urine (B), four identical samples of cow blood (C), three identical samples of cow blood (D), one sample of human blood (E), three identical samples of aqueous solution 1, and one sample of aqueous solution 2 (obtained by a 1/10 dilution of solution 1). Results for mercury are shown in Table VII. The authors found the amount of experience a laboratory had in doing an analysis had little effect on the precision and accuracy obtained.

Of the heavy metals, there is probably more current interest in lead than all others. The same chaotic problem can be shown with lead data. Lauwerys *et al.* (28) reported lead values for the samples just described. This study also involved 66 European laboratories. A selection of the lead results is given in Table VIII. Again, the results show a surprisingly large range of values, even on the aqueous samples.

An interlaboratory study on lead in a wide variety of sample types was carried out by Loescher (29). Laboratories were generally those in

Table VII

Mercury in Blood and Urine

Sample	Units	n	Mean	Range	Labs (%) with $CV_0 \leq 10\%$ [a]
Blood C	μg/100 ml	18	4.4	1.9–9.4	44
Blood D		18	2.8	0.8–11.3	33
Blood E		18	2.6	0.2–9.0	—
Urine A	μg/liter	29	7.5	1.0–88	41
Urine B		29	13.0	5–63	72

[a] CV_0 is the true interlaboratory mean calculated on replicate sample supplied.

Table VIII

Lead in Blood, Urine, and Aqueous Solutions

Sample	Units	n	Mean	Range	Labs (%) with $CV_0 \leq 10\%$[a]
Blood C	μg/100 ml	51	15	3–49	47
Blood D		52	24	10–87	65
Blood E		55	23	1–115	—
Urine A	μg/liter	33	63	5–159	42
Urine B		34	84	8–185	47
Aqueous solution 1	μg/liter	51	209	25–670	65
Aqueous solution 2		50	194	18–580	—

[a] CV_0 is the true interlaboratory mean calculated on replicate samples supplied.

the Toronto area that had been involved in the study of lead pollution around secondary lead smelters in metropolitan Toronto. A summary of some of the results is given in Table IX.

A number of agencies throughout the world supply standard reference samples (e.g., the National Bureau of Standards and the United States Geological Society in Washington, D.C. and the United States Environmental Protection Agency in Cincinnati, Ohio). It is essential that every analytical laboratory have a good selection of appropriate standards to be used for quality control.

CHOOSING INSTRUMENTATION

It is worth reiterating an earlier statement on the present state of the art in instrumentation. If one reads Walsh's original paper on atomic absorption (2) and looks again at the instrument he first publicly exhibited (Fig. 5), it is obvious that few innovations of a fundamental nature have been made since 1955. Perhaps the most important developments are technology for background correction, electrothermal atomizers, and the revolution in data handling now culminating in microprocessor-controlled instrumentation.

There are presently five major suppliers of atomic absorption instruments: Perkin-Elmer, Varian-Techtron, Instrumentation Laboratories, Fisher-Jarrel Ash, and Pye-Unicam. The prospective buyer is faced with a bewildering array of conflicting claims and optional extras.

The customer must carefully consider the type of work he wishes to do. He must be certain that this is best served by atomic absorption spectroscopy. Analyses can be categorized as follows: (1) major-

Table IX

Lead Values for Water, Sludge, and Sediment

	Water (μg/liter)			Sludge (μg/g)	Sediment (μg/g)	
	1	2	3		1	2
Mean of labs	6	4	334	1690	14	14
Standard deviation[a]	4	3	56	120	5	6
Range	2–130	2–170	280–420	1–2500	5–20	4–25

[a] Standard deviation computed after removal of outliers. Waters 1 and 2 are splits of single sample.

element, high-accuracy analyses; (2) accurate trace analyses; (3) high-precision trace analyses; (4) combinations of these. Those wishing to work in category 1 require a flame instrument. There will likely be no need for background correction accessories. A furnace is most certainly unnecessary. Workers contemplating analyses in categories 2 and 3 need background correction and may or may not require an electrothermal atomizer depending on the sample type and elements to be analyzed. In most cases, accessories can be added at a later stage. The customer must be careful not to be oversold.

Flame Equipment

In choosing flame accessories the customer should not skimp, particularly when equipment is to be used for routine operations. It is best to have highly automated gas control with all available safety features. It is also useful to purchase a nitrous oxide as well as an air–acetylene burner. Although many of the metals can be analyzed using air–acetylene flames, for a few, e.g., vanadium, silicon, aluminum, osmium, and molybdenum, the hotter nitrous oxide flame is essential. This flame gives better chemical interference suppression and is even often used for elements such as calcium and chromium.

Electrothermal Atomizers

The major commercial offerings are graphite furnace and carbon rod atomizers. In general, the graphite furnace is most satisfactory. It maximizes sample size and suffers least from reaction with the sample constituents. Reproducibility is generally best with the graphite furnace. Graphite tubes should be internally grooved to restrict sample spreading. A pyrolitic coating of the graphite gives better results for a large number of elements.

Autosampler accessories can be obtained for both furnaces and rods. These are highly recommended for improving reproducibility. A recorder is essential for work with electrothermal devices.

Double Beam versus Single Beam

Varian-Techtron has dominated the sales of single beam equipment. Hollow-cathode lamps are now very stable, thus negating the necessity of double beam optics to overcome this source of instability. Electrodeless discharge lamps, on the other hand, require 30 min warm-up periods and therefore can be used more quickly with double-beam equipment. When background correction is necessary, equipment with double-beam background correction is desirable. The deuterium arc tends to vary in intensity as do electrodeless discharge lamps. The double-beam instrument is probably a better buy when large numbers of routine analyses are to be done, particularly when nonskilled personnel are involved.

Dual-Channel Instruments

Dual channel equipment might be useful in two main instances. Two-element simultaneous analysis is possible in principle. In practice it cannot usually be done when high accuracy and precision are necessary. Double-beam background correction is possible with dual-channel equipment. Claims for superiority based on capability for internal standard operation—are generally unimpressive.

Line Sources

Conventional hollow-cathode lamps are still the most useful line sources for most elements. Boosted (high-intensity) hollow-cathode lamps may be introduced in the future for some of the more refractory elements.

Electrodeless discharge lamps are now offered commercially. These are highly recommended for elements such as arsenic, selenium, phosphorus, lead, and tin. Lamp warm-up time is about 30 min, but this disadvantage is greatly overweighed by long lifetime and higher spectral output.

Microprocessors

At present, rapid developments are being made in the area of micro-processor utilization in atomic absorption instrumentation. Some manufacturers now offer completely microprocessor-controlled instrumentation. This is a very desirable step forward. One should note, however, that atomic absorption spectroscopy is a technique based on various laws of physics and chemistry. These fundamentals dictate problems and interferences that will always be present. While automation and microprocessor control can ease the operation of instrumentation, they cannot negate spectroscopic principles. The analyst must therefore continue to be diligent in his efforts to understand basic atomic absorption theory.

Sensitivity and Detection Limit

It is important to have a means of anticipating the signal response of an atomic absorption spectrometer to a given element. This is normally done by expressing the sensitivity or detection limit of that element.

The sensitivity of an atomic absorption determination of a given element is obtained by calculating the slope of the analytical curve. On the other hand, the detection limit is that concentration of an element that gives a signal equal to some multiple, usually twice, the standard deviation of the noise. Thus of these two parameters only the latter is dependent on the S/N ratio of a determination. To obtain a detection limit figure, it is important to do at least nine determinations at concentrations near the level of the blank.

Table I contains flame detection limits obtained for the Perkin-Elmer 603 using standard water solutions (10) and Table II shows the detection limits for the HGA 2100 using the same model instrument and similar solutions.

Sensitivity and detection limit for a given element depend on many factors. The most important factors are the resonance wavelength chosen, the type of instrument, the type of atomizer, the choice and adjustments of burner, flame and nebulizer, and the type of solvent. In addition, the detection limit is dependent on the nature and amount of dissolved matrix constituents in the solvent.

It is somewhat misleading to reproduce data such as those given in Tables I and II. However, even though the numbers are not directly applicable to all atomic absorption analyses of a given element, such compilations are useful in a relative sense.

INDIRECT DETERMINATIONS BY ATOMIC ABSORPTION SPECTROSCOPY

Kirkbright and Johnson (30) reviewed indirect methods for the atomic absorption determination of elements and compounds. These methods have gained popularity for the determination of fluoride, chloride, bromide, iodide, phosphorus, and sulfur, which have principal resonance lines in the vacuum ultraviolet. In addition, indirect methods have often been employed in analyzing elements that show poor sensitivity by direct analysis. This group includes thorium, tungsten, uranium, tantalum, niobium, zirconium, rhenium, hafnium, and some of the lighter rare earths. More recently a number of indirect atomic absorption methods have been proposed for organic compounds.

Indirect techniques are prone to severe and unpredictable chemical interferences. There are better methods than atomic absorption for the above elements. No indirect methods will be given in this book.

THE ELEMENTS—SPECIAL CONSIDERATIONS

In the following section, a discussion is given of problems that are to be anticipated in work with the more troublesome elements. Specific problems encountered in a particular analytical method are outlined in the methods section.

With the exception of mercury and the hydride elements, most comments on interferences in this section relate to flame atomizers. Interferences are complex with electrothermal atomization, and the technique of standard addition, together with background correction, is widely used to compensate for such problems.

Elements having resonance lines in the wavelength range below 3000 Å are particularly susceptible to nonspecific background interference. The effect is much more serious with nonflame atomizers than with flames. In the author's experience the following elements show the greatest problems from this type of interference: silver, arsenic, gold, cadmium, cobalt, mercury, iridium, iron, manganese, nickel, lead, palladium, platinum, antimony, selenium, tin, tellurium, and zinc.

Beryllium

Even using the nitrous oxide–acetylene flame, beryllium exhibits a large number of interferences (31). Depressions of the absorbance were noted for magnesium, silicon, and aluminum. This interference can be overcome using 8-hydroxyquinoline. Enhancement interferences are obtained from many other metals. A fuel-rich nitrous oxide–acetylene flame is recommended.

Chromium

Calcium, phosphorus, and silicon can cause a depression of chromium absorption in flames (32). This problem is minimal in most applications. Interferences from iron and nickel are particularly severe. Although chromium sensitivity is best in a fuel-rich flame, interferences are greater. It is advisable to use an oxidizing flame to minimize iron and nickel interferences. Ammonium salts, e.g., 2% ammonium chloride, have been recommended to suppress interferences (33). Sodium chloride may also be used. The nitrous oxide–acetylene flame is often employed in overcoming interferences in chromium work.

Hydride-Forming Elements

The chalconides (arsenic, selenium, tellurium, antimony, bismuth and tin) and germanium form volatile covalent hydrides when they are reacted in a reducing environment in solution. The hydride thus produced is atomized in a flame or heated tube. A number of reductants have been proposed, e.g., zinc, magnesium–titanous chloride solution, and sodium borohydride solution. Most workers now use the latter.

Of the preceding elements, arsenic and selenium are particularly well handled by the hydride method. The most useful resonance lines of arsenic and selenium are 1937 and 1960 Å, respectively. At these wavelengths absorption by air and flame gases reduces the signal intensity by at least 90% for arsenic, even when using an air-entrained hydrogen flame. Detection limits obtained using conventional nebulization and normal hollow-cathode lamp sources are relatively poor.

The hydride method improves atomization efficiency greatly, by quantitatively removing the element as a hydride into the gas phase. This gas, containing most of the element, is then injected in a single pulse into a flame or hot-tube atomizer. If the latter is used, flame gas absorption is minimized and a longer residence time of atoms in the optical beam is achieved.

In this regard, a number of companies have introduced electrodeless discharge lamps (electrodeless discharge lamps have about a tenfold increase in useful intensity over conventional hollow-cathode lamps). These are to be highly recommended. Present electrodeless discharge lamps, in constrast to early models, are reasonably stable in output. Electrodeless discharge lamps cannot be powered by the hollow-cathode supply in the instrument; hence the prospective user must purchase a separate power source for these lamps. Warm-up periods of 30 min are required.

In addition to the elements already mentioned, lead forms the volatile hydride PbH_4. A few workers have suggested the hydride method for lead. However, because of the relative instability of the compound, the approach cannot be recommended.

Apparatus for Hydride Generation

Most atomic absorption instrument manufacturers offer apparatus for hydride generation as an accessory. This equipment consists of the generation vessel and a gas injection system. The collected hydrides are injected into the atomizer in a stream of inert gas, usually N_2. There have been numerous reports of homemade hydride generation equipment. Typical of this category are the zinc column generators described by Thompson and Thomerson (34) and the plastic disposable syringe used by Van Loon and Brooker (35). Automated systems for hydride generation and sample injection have been described, including a commercial offering for arsenic by Perkin-Elmer Corp.

Atomizers for Hydrides

Although some researchers have used high-temperature electrothermal graphite tubes, furnaces, etc., most atomizers in routine use are either an air–hydrogen flame or a relative low-temperature quartz tube-type furnace. Of these, the author prefers the heated quartz tube.

In using this tube it is common to burn a hydrogen diffusion flame at the open ends. Thompson and Thomerson (34) have devised a tube that allows the H_2 gas produced to burn at the ends of auxiliary tubes not in the optical beam.

Interferences

The hydride method had been used for many years, with colorimetric methods, prior to its application to atomic absorption. Hence, interference problems relating to the generation of the hydrides have been well studied. Once the hydrides have been formed, the main interference problems will be due to interaction of the covolatilized hydrides and

Table X

Interferences Observed for the Determination of the Six[a] Elements as Hydrides Generated with Sodium Borohydride in the Argon–Hydrogen Flame: Interferences Added at 1 mg Level

Element	Interference (suppression of signal)		
	Severe to moderate (50% suppression)	Moderate to slight (10–50% suppression)	Not significant (<10% suppression)
As (1 µg)	Au, Ge, Ni, Pt, Pd, Rh, Ru	Ag, Bi, Co, Cu, Sb, Sn, Te	Al, B, Ba, Be, Ca, Cd, Cr, Cs, Fe, Ga, Hf, Hg, In, Ir, K, La, Li, Mg, Mn, Mo, Na, Pb, Rb, Re, Si, Sr, Ti, Tl, V, W, Y, Zn, Zr
Bi (0.5 µg)	Ag, Au, Co, Cu, Ni, Pd, Pt, Rh, Ru, Se, Te	As, Cd, Cr, Fe, Ge, Ir, Mo, Sb, Sn	Al, B, Ba, Be, Ca, Cs, Ga, Hf, Hg, In, K, La, Li, Mg, Mn, Na, Pb, Rb, Re, Si, Sr, Ti, Tl, V, W, Y, Zn, Zr
Ge (2 µg)	As, Au, Cd, Co, Fe, Ni, Pd, Pt, Rh, Ru, Sn, Sb, Se	Bi, Cu, Ir, Te	Al, Ag, B, Ba, Be, Ca, Cr, Cs, Ga, Hf, Hg, In, K, La, Li, Mg, Mn, Mo, Na, Pb, Rb, Re, Si, Sr, Ti, Tl, V, W, Y, Zn, Zr
Sb (1 µg)	Au, Co, Ge, Ni, Pt, Pd, Rh, Ru	Ag, As, Cr, Cu, Re, Se, Sn	Al, B, Ba, Be, Bi, Ca, Cd, Cs, Fe, Ga, Hf, Hg, In, Ir, K, La, Li, Mg, Mn, Mo, Na, Pb, Rb, Si, Sr, Te, Ti, Tl, V, W, Y, Zn, Zr
Se (2 µg)	Ag, Cu, Ni, Pd, Pt, Rh, Ru, Sn	Au, As, Cd, Co, Fe, Ge, Pb, Sb, Zn	Al, B, Ba, Be, Bi, Ca, Cr, Cs, Ga, Hf, Hg, In, Ir, K, La, Li, Mg, Mn, Mo, Na, Rb, Re, Si, Sr, Ti, Tl, V, W, Y, Zr (Te)
Te (10 µg)	Ag, Au, Cd, Co, Cu, Fe, Ge, In, Ni, Pb, Pd, Pt, Re, Rh, Ru, Se, Sn, Te	As, Bi, Ir, Mo, Sb, Si, W	Al, B, Ba, Be, Ca, Cr, Cs, Ga, Hf, Hg, K, La, Li, Mg, Mn, Na, Rb, Sr, Ti, V, Y, Zn, Zr

[a] Results for tin are not quoted because of the high blank.

nonspecific absorption effects. Smith (36) presents a comprehensive review of interference problems (Table X) and concludes that they are generally of the two types previously indicated.

Elements that may be expected to interfere with the generation of hydrides are those such as copper, silver, gold, and iron, which consume reductant in competition with the element of interest. This problem is not serious as long as an excess of the reducing reagent is used. It is obvious from the author's work, however, that the period required to obtain quantitative generation of the hydride is very dependent on the matrix. Hence, it is important to experiment with actual matrix samples to determine the proper reaction time required in a specific application.

Determination Directly with an Electrothermal Atomizer

Several of the base metals, e.g., iron and nickel, stabilize arsenic, selenium, or tellurium during thermal pretreatment (ashing). This allows the determination of these elements without fear of loss as volatile compounds prior to atomization. In some samples the base metals are naturally present; in others, such as biological materials, they must be added.

There have been a number of procedures published for the direct determination of these elements using conventional electrothermal atomization. The author believes that for arsenic and selenium the hydride method is definitely superior. For tellurium electrothermal methods can readily be employed. Too little work has been done on the other elements to give a definitive answer as to the relative merits of these approaches at this time. The analyst must bear in mind, however, that the hydride method has the advantage of removing analyte elements from many potential interferents.

Iron

Few problems in the flame atomic absorption determination of iron have been reported. A highly oxidizing air–acetylene flame is best (37). The iron spectrum is particularly rich in absorbing lines. The most sensitive wavelength, 2483 Å, is closely surrounded by nonabsorbing lines and hence a slit width of no greater than 2 Å should be employed (37).

Lead

Lead is of particular interest to environmental and health scientists, as evidenced by the large number of publications dealing with the atomic absorption determination of this metal in body fluids and tissue and in environmental samples. Due to this activity, more information exists on the problems associated with lead analysis than is obtainable for most elements. A great deal of the latter information describes errors resulting from problems with sample preparation.

For example, Baily and Kilroe-Smith (38) studied six sample preparation methods for blood analyzed by graphite tube atomic absorption:

(1) Blood diluted five times with water.

(2) Blood diluted five times with 0.01 M hydrochloric acid.

(3) Blood diluted five times with 2% Triton-X.

(4) Blood treated with Unisol, allowed to stand overnight, and diluted by a factor of 5 with water.

(5) One volume of blood was treated with an equal volume of a mixture of perchloric acid–trichloroacetic acid, extracted for 1 hr at room temperature, and diluted with eight volumes of 0.1 M hydrochloric acid.

(6) One volume of heparinized blood was treated in an acid-washed tube with an equal volume of 70% perchloric acid. The sample was vibrator mixed and one volume of 30% hydrogen peroxide added. The mixed sample was heated in a water bath for 30 to 60 min at 60 to 80°C with occasional agitation. The sample was diluted with seven volumes of water.

The analyte plus background and background absorptions were monitored at 2833 and 2875 Å, respectively. Results showed that with methods 1–5, where the protein is not precipitated or digested, high background absorption occurs. This results in higher relative standard deviation of the results. (It should also be noted that, as described in the section on electrothermal atomizers, background correctors may fail to function properly at high background absorption levels.) Baily and Kilroe-Smith (38) favor method 5 because of procedural simplicity compared to method 6. The latter two methods gave equally good results.

The author compared methods 1, 3, and 6 and found method 1, if used as indicated on p. 174, to be satisfactory. Addition of reagents, as outlined in methods 2–6, may result in unacceptable contamination of the sample.

Few chemical interference problems have been reported for the flame atomic absorption determination of lead. Magnesium and aluminum cause problems when their concentrations are greater than 2.5% (39).

There are many references in the literature to the loss of lead during ashing. Most authors recommend an ashing temperature of less than 400°C. Work in the author's lab suggests that higher ashing temperatures are permissible in the presence of the oxidizing acids. Workers are cautioned, however, to check for losses on each new sample type before using an ashing temperature for routine analysis. This can be done as suggested on p. 296.

The 2833 Å lead line, although less sensitive by a factor of 2 than the 2170 Å line, is to be strongly recommended. This is due to its better source signal strength and the lower level of background interference experienced at this wavelength. Despite this observation, it is essential to employ background correction for most lead analyses using flameless atomizers.

Sampling cup and boat methods for analyzing lead in biological samples are common. Most workers agree that when an electrothermal furnace type atomizer is available, this should be used in preference to sampling cups.

Mercury

In 1930, Müller (1) used atomic absorption for the analysis of mercury in air. Light from an ultraviolet lamp was passed through an absorption cell, the absorption being monitored by a photometric readout. Since the Minamata mercury incidents in Japan and the recent interest in mercury in the earth sciences, atomic absorption methods for the determination of mercury have been the subject of extensive investigation.

Mercury is present in most samples at sub-ppm levels. The sensitivity obtained by nebulizing a mercury solution into a flame is seldom satisfactory. The problem is compounded by the fact that the most sensitive resonance line is at 1849 Å. This wavelength cannot be used with conventional atomic absorption units. The commonly employed 2537 Å line is 40 times less sensitive. Because metallic mercury has an appreciable vapor pressure at room temperature, a cold vapor containing atomic mercury is produced from the sample and swept, in a stream of air, into the absorption cell. The cell is maintained at room temperature. The residence time of the atoms in the optical beam is longer compared to a flame, thus greatly improving the sensitivity.

Sample Drying

A good deal of controversy exists about loss of mercury during the drying of samples. The magnitude of this loss depends on the temperature used, the form of mercury in the sample, and the sample type. Volatile forms of mercury—the metal and organic compounds—are readily lost even at slightly elevated temperatures. It has been shown that freeze-drying and vacuum desiccation result in losses of mercury in organic forms.

Because of these problems, drying at room temperature has become the most generally accepted procedure. However, it must be recognized that even under this condition, loss of mercury can occur.

It is often important to study mercury loss due to drying. A procedure for doing this is given on p. 296 in the section on selective volatilization of metals and metal compounds.

Sample Preservation

The mercury content of water samples decreases on standing. This may be due to a number of processes including adsorption, bacterial action, and reduction by organic matter. Several methods have been recommended to overcome this problem. Two of these appear to be most generally acceptable, stabilization with oxidant or with gold.

Most workers use an oxidant in acid solution as a preservative. The samples are filtered through 0.45 μm porosity membranes, acidified to about 0.5 M HNO_3 and made 0.05% in potassium dichromate (40). An acidic solution of potassium permanganate is also often used (40). Recently it has been shown that the addition of gold stabilizes mercury in water. The sample is made 0.5 M in HNO_3 and 10 μg/ml in gold (41).

There is some evidence to suggest that solid samples can also decrease in mercury content with time. To minimize this possibility, samples may be stored in a refrigerator.

Sample Decomposition

A wide selection of sample decomposition procedures exists. These are basically of two types: a pyrolysis or an oxidizing decomposition in an acid medium.

Considering the acid decomposition approach, a number of alternatives exist. To avoid loss of mercury, the decomposition must be kept oxidizing at all times. It is common to employ nitric acid or nitric and sulfuric acids with potassium permanganate or potassium persulfate. Some workers have used perchloric acid with nitric acid.

Pyrolitic decomposition is usually accompanied by an amalgamation separation and preconcentration (42). This allows the separation of mercury from smokes and organic vapors, which can cause serious interference. Gold is the most commonly used metal for amalgamation traps.

Several manufacturers offer a mercury analysis kit to be used with atomic absorption, based on the approach described by Hatch and Ott

(17). This apparatus usually consists of a glass decomposition vessel, a recirculation pump, and an absorption cell.

Interferences

Nonspecific absorption interference can be a serious problem following both pyrolitic and acid decomposition methods. Organic vapors, inorganic molecular gases, smokes, and mists from the sample solution can all contribute to the problem. For this reason it is best to use an atomic absorption unit equipped with a background corrector.

A number of anions have been reported to depress the absorption of mercury by suppressing its evolution from solution. These include bromide, nitrate, perchlorate, phosphate, sulfide, and sulfate. Cations of the nobler metals, e.g., gold, platinum, palladium, silver, and copper, cause a similar problem. Selenium and tellurium can also interfere. To minimize interference due to anions, it is best to use a similar matrix composition medium for both samples and standards.

Molybdenum

Interference patterns for the flame atomic absorption analysis of molybdenum are very complex (20, 43). Depending on the anion present, a given cation may show both an enhancement and depression (20). For example, sodium as the sulfate or nitrate gives a signal enhancement, but a signal depression occurs when sodium is present as the chloride. Other commonly occurring cations that cause severe interference problems are aluminum, potassium, calcium, magnesium, and iron. A nitrous oxide flame is recommended. Samples and standards should be prepared to contain 1000 ppm aluminum to minimize interference problems.

Nickel

The most sensitive line is at 2320 Å. However, a number of nonabsorbing lines lie close enough to this wavelength to be potentially within the bandpass of the optical system. Although this is not an interference problem, nonlinear calibration curves and poor sensitivity will result if nonabsorbing lines are not rejected. It is important, therefore, to use as small a slit width (1 Å or less) as is consistent with a usable S/N ratio. When background correction employing a continuum source is necessary, narrow slit widths are difficult to maintain. This is due to the greatly reduced signal of the continuum at small slit widths.

Chemical interferences may be a problem in nickel determinations in flames. Sundberg (44) and others recommend an oxidizing flame to minimize this effect. Work done by Sundberg with 2000 μg/ml solutions of Fe(III), Mn(II), CR(III), Cu(II), Co(II), and Zn(II) containing 20 μg/ml nickel demonstrates that the nature of the matrix determines the height in the flame at which maximum concentration of the ground-state nickel atoms occur. This type of problem has also been noted for other elements. It has been termed a lateral diffusion interference.

Noble Metals

Noble metals have traditionally presented the ultimate challenge to the inorganic analytical chemist. This remains the case in the application of atomic absorption to the determination of these elements.

Strasheim and Wessels (45) in 1963 were the first workers to describe, in some detail, the complexity of interference patterns associated with analyses in base-metal sample solutions. Their work was largely ignored by analysts in early years, as is evidenced by the numbers of erroneous results generated in routine applications. However, during the past decade a number of analytical researchers have studied the interference problems and popularized useful remedies. Currently atomic absorption is the preeminent determinative tool for the noble metals. The truth of this statement is strikingly illustrated by the fact that in a 1975 certification study of a noble-metal ore sample (46), 77% of all the results reported were obtained by atomic absorption.

The detection limits of the noble metals by atomic absorption are not low enough for their direct determination in rocks and ores. In general, it can be stated that gold, platinum, rhodium, and palladium have useful detection limits such that a preconcentration step easily brings them into the working range. Iridium, osmium, and ruthenium exhibit poor sensitivity. In spite of this it is common to employ atomic absorption for the analysis of these elements. This is particularly true in the case of iridium. The author does not recommend atomic absorption for the determination of osmium and ruthenium. These elements can be done with ease by the thiourea spectrophotometric method, following a distillation separation. Despite this, methods for osmium and ruthenium are included for workers who are unable to use the thiourea method.

There are two highly recommended preconcentration/separation approaches for the noble metals prior to atomic absorption analysis: fire assay and solvent extraction. Sometimes both are used. Fire assay is so highly revered that 92% of all the results reported in the 1975 certification study (46) were obtained by this technique.

Most researchers believe it to be a mistake not to use fire assay preconcentration for any rock or ore. This stems from two problems. First, noble metals, particularly gold, are commonly very inhomogeneously distributed in these samples. This necessitates the use of a large sample (10–100 g). Second, noble metals are very resistant to attack by mineral acids, particularly when entrapped in a particle of siliceous gangue. Fire assay breaks down all siliceous material.

Of the available fire assay procedures, the classical lead assay is most generally preferred. It can be used, followed by cupellation into a silver bead, for platinum, palladium, rhodium, and gold. When iridium is to be done, cupellation into a gold bead is satisfactory. Other assay procedures such as the iron–copper–nickel, nickel sulfide, and tin methods can be used, but analysis in the resultant sample solution is more difficult.

Solvent extraction can be used, in some instances, for the recovery of noble metals from acid digests of ores. However, as indicated above, the efficiency of such procedures for a given sample type must be first checked by fire assay. More commonly, solvent extraction is used to recover noble metals from samples such as cyanide plating baths and dissolved alloys and other commercial products.

Interferences

In early work, Lockyer and Hames (47) and Menzies (48) reported no interference in the determination of platinum, palladium, rhodium, and gold. Strashiem and Wessels (45) recorded interference from other noble metals and base metals in the determination of platinum and rhodium. The problem, associated with the other noble metals, was overcome with 20,000 ppm copper as sulfate.

In the past decade a surprising variety of releasing agents and buffers have been proposed to remedy interference problems.

Pitts *et al.* (49), Van Loon (50), and Schnepfe and Grimaldi (51) used lanthanum to eliminate a wide variety of base metal and noble-metal interferences in the determination of platinum, palladium, gold, and rhodium. For the determination of iridium, Grimaldi and Schnepfe (52) and Van Loon (53) used a mixture of copper and sodium as an interference suppressant. Interferences in the determination of rhodium, platinum, palladium, iridium, and gold were minimized using copper and cadmium in a study by Sen Gupta (54). Mallet *et al.* (55), Jansen and Umland (56), and Mallet and Breckenridge (57) found that vanadium, uranium and vanadium, or uranium, respectively, could be used to suppress interference in the determination of the noble metals in a variety of complex matrix solutions. All the above researchers used air–acetylene flames.

A number of authors, e.g., Pitts and Beamish (58), have found that the hotter flame, nitrous oxide–acetylene, reduces interferences significantly. However, the noise levels are appreciably higher, making detection limits poorer.

Some work, e.g., by Pearton and Mallett (59) and Everett (60), has been done to evaluate the applicability of electrothermal atomizers to the analysis of noble metals. Results are preliminary at present and the technique has not received enough attention to allow its routine application at this time.

Tin

Tin absorbance in the air–hydrogen flame is three times higher than for the air–acetylene flame. However, the interference patterns are much worse. It is best, according to recent literature (61), to use the nitrous oxide–acetylene flame to minimize interferences. As already mentioned, tin can also be analyzed by the hydride method.

Vanadium

Interferences are complex in the determination of vanadium. A strongly reducing nitrous oxide acetylene flame is to be recommended.

To minimize interferences in the nitrous oxide–acetylene flame, fuel-rich (highly reducing) conditions must be used (62). The addition of 1000 μg/ml aluminum to standards and samples overcomes residual interference problems (63). If an air–acetylene flame must be used, 1000 μg/ml aluminum should be added to overcome interferences (63). Titanium (64) has also been used for this purpose.

Zinc

The flame atomic absorption determination of zinc is relatively free from chemical interferences. Silicon is the only significant problem (65).

The most sensitive line for zinc is at 2139 Å. At this wavelength about 25% of the incident radiation is absorbed by a standard air–acetylene flame (37). It is important, therefore, to maintain a constant flame composition throughout the analysis.

Nonspecific background absorption is particularly severe in the determination of zinc using electrothermal atomizers. This, combined with the ubiquitous nature of zinc, makes its determination at very low levels extremely difficult.

PREPARATION AND STORAGE OF SOLUTIONS

Water

High-purity water for atomic absorption purposes can be produced by distillation or with ion exchange resins, or with a combination of both. The analyte element and the concentration at which this element must be analyzed will determine the quality of water required. Ubiquitous elements such as zinc, calcium, potassium, sodium, and iron, when they must be analyzed at submicrogram/milliliter levels, dictate extreme precautions in the production and subsequent use of high-purity water. In many laboratories, including that of the author, it is impossible to analyze these elements below about 1–5 μg/liter because of uncontrollable contamination problems.

Ion Exchange Systems

Depending on the characteristics of the water supply and the purity requirements of the finished water, one or several stages are used within an ion exchange system. Incoming water is first filtered to remove particulate material. If organic matter is a potential problem the water can be run through an activated-charcoal column. Finally, the water is passed through one or more stages of a mixed cation–anion exchange column.

Cation and anion resins are used in H^+ and OH^- forms, respectively. Contaminant ions in the water supply exchange onto the column displacing H^+ and OH^-, which combine to form water.

Distillation

For many purposes, one stage of distillation using commercially available equipment is sufficient. In general, particularly when metal analyses are to be done, glass or quartz apparatus should be used.

A variety of organic contaminants distill over with the water. To negate this problem, alkaline potassium permanganate can be added to the distillation flask. In the presence of permanganate most offending organic substances are oxidized to carbon dioxide and water.

When high-purity water is essential, two stages of distillation are often employed. Again, at least the final stage should be accomplished using an all-glass system. Double distillation may also be essential when the water supply is high in total dissolved solids. In this case a combination of ion exchange and distillation is a good approach.

For the ultimate in high-purity water, a sub-boiling distillation should be employed. As would be expected, this is extremely slow and expensive ($<$ 1 liter/day) and hence should be used only when ultrapure water is a necessity. For storage of this water, a Teflon bottle minimizes contamination.

Standard Solutions

Great care is, of course, required in the production of standard solutions. These will be used to prepare calibration graphs and hence will in large part determine the accuracy of an analysis. Standard reference samples must also be used to assure accuracy. These samples are often instrumental in pointing up errors in standard solutions.

Reagents

For most elements, the high-purity metal is the substance of choice for preparing standard solutions. In the majority of cases a metal of at least 99.9% purity is sufficient. The analysis of alloys or high-purity metals may require addition of the major matrix element to the standard solutions to suppress chemical interference. In this case metals of ultrahigh purity must be obtained and checked for absence of the analyte element.

The metals of some elements are not suitably stable or cannot be obtained in a sufficiently pure state. In these instances a high-purity salt may be employed. The salt chosen should not be hygroscopic and should be of known composition. High-purity metals and metal salts are available from several companies. (The author can suggest Spex Industries, Box 798, Metuchen, New Jersey; Johnson Mattley Chemicals, 74 Hutton Garden,London ECIP 1AE, England; and Fisher Reagent Grade Chemicals, Fisher Scientific, Fairlawn, New Jersey.) Table XI contains suggestions of metals/metal salts and solvents for preparation of standard solutions. In each case, when acid is used, the final acid concentration in the diluted solution should be about 1%. Further details on standard solutions are given with the procedures in the following chapters.

Reagents other than metals required for standard solutions, such as acids, buffers, and releasing agents, must be of high purity. A blank of these reagents should always be prepared.

There are commercially available atomic absorption standard solutions. These are satisfactory for many applications.

Containers

There is no unanimity of opinion on the best type of container to use for the storage of standard solutions. The controversy surrounds the reported loss or gain of metals resulting from exchange and adsorption on container walls. Choices range from various glass vessels to a wide variety of plastic containers.

With the possible exception of mercury, the author recommends high-density, linear polyethylene for storage of standard solutions. This material when properly washed yields minimal metal contamination and does not suffer from the severe evaporation (breathing) problems encountered with other plastics. Teflon containers, used by the National Bureau of Standards for reference water samples, are superior. The cost of these vessels negates their use for the storage of the large numbers of standard samples employed by most laboratories.

Glass containers are inferior to plastics for most metals because of a higher incidence of ion exchange and adsorption problems on the walls. The possible exception to this rule is the storage of mercury solutions.

Plastic containers, as supplied, are often contaminated with traces of metal, particularly zinc. A rigorous regime of decontamination is thus required. Most workers recommend a detergent wash followed by an acid treatment and distilled water rinses. The author agrees with this approach, but recommends against the use of hot concentrated acids for the acid wash. The latter may attack the surface of the plastic. Soaking 0.5 hr in a room temperature, 1:1 mixture of hydrochloric or nitric acid followed by several rinsings with the same mixture is best. Distilled water should be used for all water rinsings.

Preparation of Standard Solutions

The metal or metal salt should be dried at an appropriate temperature and cooled in a desiccator prior to weighing. The latter must be done to about 1 part per thousand for most applications. Stock standard solutions are usually made to contain 1000 $\mu g/ml$.

The weighed product should be dissolved in a minimum of acid. Hydrochloric or nitric acid should be used when possible, bearing in mind chemical problems that might be encountered (e.g., precipitation of silver chloride in the presence of hydrochloric acid). Nitric acid is preferred when the solutions are to be used in an electrothermal atomizer. The solution must be diluted to volume in acid washed, class

Table XI

Suggested Metal/Metal Salt and Solvent For Preparation of Standard Solutions

Element	Metal/metal salt	Solvent	Comments
Ag	Silver nitrate	Water	Store in amber bottle
Al	Aluminum metal	1:1 HCl	Add a drop of mercury to aid dissolution
As	Arsenious oxide	20% KOH	Follow procedure on p. 60
Au	Gold metal	Aqua regia	Store in amber glass bottle
Be	Beryllium metal	1:1 HCl	Metal dust is very toxic if inhaled
Bi	Bismuth metal	1:1 HNO_3	
Ca	Calcium carbonate	1:3 HCl	Add dropwise
Cd	Cadmium metal	1:1 HCl	
Co	Cobalt metal	1:1 HCl	
Cr	Potassium chromate	Water	
Cu	Copper metal	1:1 HNO_3	
Fe	Iron metal	1:1 HNO_3	
Hg	Mercuric oxide	1:2 HCl	
K	Potassium chloride	Water	
Mg	Magnesium metal	1:1 HCl	
Mn	Manganese metal	1:1 HNO_3	
Mo	Ammonium paramolybdate	1% NH_3	Final concentration of ammonia in the diluted solution should be 1%

	Compound	Reagent	Notes
Na	Sodium chloride	Water	
Os	Ammonium chloro-osmate	1% HCl	Osmium is very toxic and very volatile as the tetroxide
Pb	Lead metal	1:4 HNO_3	This is best but an oxide coating may form on the metal, making dissolution troublesome
Pd	Lead nitrate	1% HNO_3	
Pt	Palladium metal	Aqua regia	
Rh	Platinum metal	Aqua regia	
Ru	Ammonium hexachlororhodate	10% HCl	Diluted solution should be 10% in HCl
Sb	Ruthenium chloride, $RuCl_3$	20%	Diluted solution should be 10% in HCl
Se	Potassium antimony tartrate	Water	
Si	Selenium metal	Concentrated HNO_3	Evaporate to dryness; add water and evaporate to dryness; dissolve in 10% HCl; diluted solution should be 10% in HCl
Te	Silicon dioxide	Lithium metaborate fusion	Fuse with sevenfold excess of lithium metaborate and equal weight of calcium oxide in a preignited graphite crucible at 925°; pour melt into 1:24 HNO_3; dilute with 1:24 HNO_3
Zn	Tellurium dioxide	Aqua regia	Dilute with 6 N HCl
	Zinc metal	1:1 HNO_3	

A or equivalent, volumetric flasks. This dilution and subsequent dilutions should be made so that the final acid concentration is about 1%. Maintaining a 1% acid concentration in the standard minimizes hydrolysis and ion exchange and adsorption problems on container walls. For some elements, e.g., sodium, potassium, and calcium, it is not necessary to maintain a 1% acid content. These elements are not prone to hydrolysis. Suggested dissolution approaches are given in Table XI.

Preservatives may be necessary for a few elements. For example, in the case of the mercury a few microgram/milliliter of gold or 0.05% potassium dichromate yields more stable solutions.

2
Analysis of Waters

The analysis of water would seem, at first glance, to be the simplest of all chemical analyses. This common assumption has lead to a shocking amount of erroneous trace metal data.

Water samples span a wide variety of matrices. Sea and brackish waters and some effluents contain high levels of dissolved salts. In the case of effluents, the main matrix constituents can vary widely from sample to sample. Waters with high dissolved-salt content must, in most cases, be treated by solvent extraction or ion exchange to remove the desired trace constituents from the matrix prior to analysis. The determinative step can then be accomplished by flame or electrothermal atomization.

Contrary to much published material, the author cannot recommend the direct analysis of high-salt-content waters using electrothermal devices. Despite skillful manipulations during drying and ashing, including the addition of reagents, the magnitude of the background signal on atomization often runs close to or exceeds the capability of the background correction system. Good solvent extraction–flame methods exist for samples of this type.

When dealing with "soft" waters (rain, lakes, rivers, etc.) it is often possible to analyze samples directly with the electrothermal atomizers. As a precaution the author runs a few samples by extraction to check on the reliability of these results. Evaporation as a method of preconcentrating elements in waters is not recommended because of the contamination hazards.

Gibbs (66) identified five mechanisms by which metal can be transported in waters: in solution and organic complexes, adsorbed, precipitated, coprecipitated in organic solids, and coprecipitated in crystalline sediment material. Whether one agrees with such a classification or not, it is obvious that metals in some forms are more available for reaction with biological components of the environment than others. It would be impossible to give a meaningful discussion of metal availability in a book of this nature, mainly because of the wide range of opinion that exists. In spite of this it is probably useful to distinguish between "easily extractable" metal and "total" metal in waters. Extractable metal in this context is the metal, both bound and unbound, that can be extracted by the organic reagents used according to a proposed procedure. Total metal refers to the metal released by the acid digestion of the sample. This metal is termed "total" because the author's experience indicates that over 90% of the metal in the water is released by acid digestion.

No general accord exists on the method of water pretreatment prior to analysis. Most authors agree on an acidification following filtration. The filtration should be done at the time of sample collection or within a few hours thereafter. It is common to filter waters through a 0.45 μm porosity filter. Metal that passes this porosity is often termed soluble. The author believes that a significant percentage of particulate metal passes this filter porosity size. Work designed to clarify some of the above variables is in progress in the author's laboratory, embracing a wide range of water sample types.

As previously indicated in the section on standard solutions, a good deal of controversy exists on what type of bottle should be used for storage of water samples. Choices range through various glasses and plastics. The author has found that high-density linear polyethylene is a good inexpensive container for water samples. Generally, glass vessels are a poor choice because of problems resulting from ion exchange on the walls. Plastics tend to "breathe" and hence, due to evaporation, metal levels become higher. High-density linear polyethylene exhibits some problems due to evaporation but is a good compromise between impervious, expensive Teflon and very porous soft polyethylene.

All containers for water samples must be acid washed and rinsed in distilled water. Generally, it is best to use a 1:1 acid strength (nitric acid is good) for this purpose.

SOLVENT EXTRACTION FOR THE PRECONCENTRATION OF METALS IN WATERS

Ion exchange chromatography and solvent extraction are the most frequently used methods for preconcentration and/or separation of metals from matrix constituents in waters. In the author's laboratory, solvent extraction has been found to be the most satisfactory approach. Apart from yielding a satisfactory separation, solvent extraction has the advantage that the metals of interest are in an organic solvent. If the proper solvent is chosen, an enhanced signal compared to water solutions is obtained when flame atomization is employed.

Solvent extraction methods for concentrating trace metal ions in waters, prior to atomic absorption analysis, abound. Unfortunately, the majority of these have been developed without regard for important theoretical data available from such sources as Stary (67), Morrison and Freiser (68), and Zolotov (69). There have only been a few critical studies of solvent extraction–atomic absorption procedures. As a result, available procedures are seldom optimized with respect to pH range, buffer, ionic strength, stability, equilibration time, etc. This means it is often impossible for the analyst in a laboratory to obtain good results on a routine basis.

Proposed solvent extraction methods for the atomic absorption analysis of trace metals were examined in the author's laboratory (70). The following is a summary of the important considerations that resulted.

The solvent used to extract metal complexes must have a number of desirable characteristics. It must (1) extract the desired metal chelates, (2) be immiscible with the aqueous solution, (3) not tend to form emulsions, (4) have good burning characteristics, and (5) enhance rather than suppress the atomic absorption sensitivity as compared to the metal in water.

Work was carried out on a number of likely solvents. Benzene and xylene were eliminated because of the turbulent and unstable flames they produced. Decanol proved to have too pungent an odor. Chloroform, a solvent widely used in colorimetric work, evaporates too quickly, leaving the solid complex behind, which then clogs the vaporization chamber (71). Ethyl acetate, methyl isobutyl ketone, isoamyl acetate, and N,N-butyl acetate were also considered. Of these, ethyl acetate and methyl isobutyl ketone gave the greatest enhancements as compared to the absorbance of the same quantity of metal in water. Ethyl acetate is too volatile for easy use. Although methyl isobutyl ketone is fairly soluble in water, this extractant was chosen.

The use of a buffer is mandatory in routine extraction work. This fact is not recognized by many workers. It is well known that the quantity of metal extracted is strongly dependent on the pH of the solution and that chelating agents will often alter the pH of the solution to which they are added.

The choice of the buffer is very important. It must be stable, have a high buffering capacity, and not participate in any reaction. A number of buffers have been studied for solvent extraction–atomic absorption work: borate, phosphate, citrate, acetate, and formate. Solutions containing the formate buffer were found to be unstable and slowly decomposed (organic droplets appeared on container walls) after several days. An acetate buffer was also unfavorable, because it would combine with any lead or silver in solution to form stable acetates, which were not readily extracted. A citrate buffer, which was found to be stable and did not interfere with the extraction process, is best for most metals. This buffer, however, does contain considerable cadmium and iron. The buffer should be purified as well as possible of trace metal contaminants by an extraction wash using the chelating agent. In spite of this precaution, a blank must be run with each set of samples. If silver and lead are not to be analyzed, an acetate buffer is recommended because of lower cadmium and iron contamination.

Two of the main considerations affecting the choice of a chelating agent are that it should (1) extract the largest number of trace metals and (2) extract the metals equally well over some fairly wide range of pH of the solution. Many procedures available at present require pH adjustment to within one pH unit or less, which can result in serious errors in routine applications.

The pH dependence of nine different chelating agents was studied. These chelating agents were 8-quinolinol, acetylacetone, thenoyltrifluoroacetone, 1-(2-pyridylazo)-2-naphthol, ammonium pyrolidinedithiocarbamate, potassium ethyl zanthate, α-benzionoxime, diethylammonium diethyl-dithiocarbamate, and sodium diethyl-dithiocarbamate. Ammonium pyrrolidine dithiocarbamate extracts the greatest number of metals over the widest pH range. Diethylammonium diethyldithiocarbamate has a stabilizing effect on all metal complexes in the system and hence its addition is recommended.

Solvent Extraction Method for the Analysis of Metals in Waters

Comment on the Method

In the analysis of water samples, a number of ionic species may be encountered at concentration levels far greater than those of the metals being determined. Fluoride, calcium, potassium, magnesium, sodium, phosphate, silicate, and biodegradable detergent are potential interferences present in many waters. These were individually added to solutions and the absorbances obtained compared to the absorbances of solutions without interferences. Most of these ions are known masking agents; however, only the biodegradable detergent appeared to affect the absorbances of the metals to any degree.

The following procedure is suitable for the analysis of sea, brackish, and fresh waters and for most effluents. Elements that can be readily determined are silver, cadmium, cobalt, copper, iron, nickel, lead, and zinc, with detection limits of 0.6, 0.8, 1.5, 0.8, 1.3, 2.5, and 0.6 μg/liter, respectively.

Reagents and Equipment

An IL 153 (Instrumentation Laboratories, Inc.) was used for all atomic absorption measurements. The pH measurements were made using a Beckman expandomatic pH meter.

All chemicals used were of reagent grade of the highest quality available. Standard metal solutions (1000 μg/ml) were prepared for Ag(I), Co(II), Fe(III), Mn(II), Ni(II), Cu(II), Pb(II), Cd(II), and Zn(II) from high-purity metal. Buffer was prepared by mixing the appropriate amounts of citric, boric, and phosphoric acids in water to give a solution that is 0.5 M in each acid.

Prepare a 1% w/v mixed chelating solution of ammonium pyrrolidine dithiocarbamate and diethylammonium diethyldithiocarbamate in water. Extract this twice with methylisobutyl ketone.

Procedure: Extractable Metals

Acidify the water to pH 1 with nitric acid. Place the desired sample (usually 200 ml) in a 250 ml separatory funnel fitted with a Teflon stopcock. Add 4 ml of the buffer. Shake to mix well. The pH should be 4.0 $\pm$ 0.1. If the pH must be adjusted, add sufficient 20% sodium hydroxide solution to obtain this value. Add 5 ml of 1% mixed chelating agent. Shake briefly. Add 10–20 ml (depending on concentration factor required) of MIBK. Shake vigorously for 60 sec. Allow the layers to separate. Remove aqueous lower layer. Retain the MIBK layer in tightly capped glass bottles until samples have been made ready for analysis. Prepare standards from multielement stock solution so that the 200 ml of water extracted contains four concentrations within the ranges iron 10–20 μg/liter, copper 5–100 μg/liter, nickel 5–100 μg/liter, cadmium 1–20 μg/liter, zinc 10–2000 μg/liter, and lead 10–2000 μg/liter. In this way a direct concentration relationship exists with samples. Run a reagent blank.

Procedure: Total Metal

Add 1 ml of nitric acid to the desired sample. Evaporate the sample to dryness on medium heat on a hot plate. Add 2 ml of hydrochloric acid and 1 ml of nitric acids. Evaporate to dryness. Add one drop of hydrochloric acid and dilute to 200 ml. Continue as for extractable metal above. Run a blank containing all reagents.

Determination of Mercury in Water (72)

Comment on Method

A cold trap preconcentration step is used in which a glass U tube is packed with glass beads and immersed in liquid nitrogen. This avoids the use of slower amalgamation or extraction preconcentration steps. The procedure was tested on sea water, marine zooplankton, NBS orchard leaves, and oceanic manganese crustal materials. The value obtained for NBS orchard leaves was 150 ± 17 ng/g, compared to an accepted value of 155 ± 15 ng/g.

Reagents and Equipment

The mercury analyses were conducted using a Coleman Instrument mercury analyzer (MAS 50) equipped with a Leeds & Northrup Speedomax Recorder (Model XL 601). The aqueous sample solution was contained in a 250 ml Pyrex (Corning Glass Works) glass bubbler placed at one end of a sampling train employing nitrogen as the purging and carrier gas. The gas flow system also included a flow regulator, two bypass valves, water absorber, a mercury cold-trap, a gas cell, and a gas washing bottle containing a 10% solution of potassium permanganate. A schematic diagram of the entire system is shown in Fig. 27.

The sparging vessel was connected directly through amber latex rubber tubing (i.d. 3 mm, length 60 cm) to a Pyrex glass drying tube (i.d. 18 mm, length 18 cm) containing colorless silica gel (6/20 mesh) as a water absorbent. The mercury cold trap followed this absorbing stage, and connection was made using a 75 cm length of rubber tubing. The mercury cold trap consisted of a 50 cm Pyrex glass (i.d. 4 mm) U tube

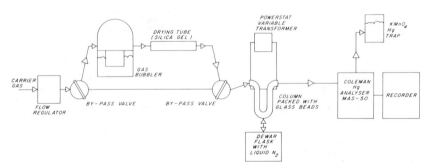

Fig. 27. Apparatus for the determination of mercury in water (72).

(width 4.4 cm), which was packed with glass beads (90/100 mesh) to form a 6 cm column at the bend of the tube. The U tube was wrapped with nichrome wire (diameter 0.06 mm) yielding five windings over the entire lower 15 cm of the tube. The wire-wound U tube was placed inside an insulating Pyrex glass covering (i.e., i.d. 10 mm, length 3.8 mm) such that only the non-wire-wrapped sections were exposed. This complete U tube apparatus was designed to fit a 1 liter Dewar flask containing liquid nitrogen to provide the trap and concentration step for mercury vapor. The leads from the heating wire were connected to a Powerstat variable transformer, which allowed the column to be heated electrically upon removal from the liquid nitrogen bath. Three-way Teflon stopcocks (4 mm bore) were placed before the bubbler and after the drying tube to permit the sparging vessel and the drying column to be bypassed during the heating and elution step. Connections were made with 3 mm i.d. rubber latex tubing. The mercury, which is rapidly vaporized and eluted from the column during the heating step, was fed directly by polyethylene tubing (i.e., i.d. 3 mm, length 70 cm) to the gas cell of the Coleman mercury analyzer. The absorption of elemental mercury in arbitrary units was displayed on a recorder using a $25\times$ scale expansion. After the carrier gas had passed through the gas cell, it was directed into a 300 ml gas washing bottle containing 100 ml of 10% potassium permanganate. At this stage, the elemental mercury was oxidized and removed from the gas flow.

Except where noted, all chemicals used were ACS certified reagent grade quality supplied by the J. T. Baker Chemical Company. The nitrogen carrier gas was 99.8% pure (Chemetron Corp.). Deionized water for reagent preparation and experimental studies was produced by passing Pyrex glass distilled water through an activated charcoal absorbent and two mixed-bed ion exchangers (Continental Dionization Service). A primary standard mercury solution was made by dissolving 1.354 g of freeze-dried spectrographic-grade mercuric chloride (Johnson, Matthey Chemical Ltd.) in 1000 ml of distilled deionized water. From this solution, 10 and 100 μg/liter spiking solutions were made daily. A 20% stannous chloride solution was prepared by dissolving 72 g stannous chloride in 20 ml concentrated hydrochloric acid and bringing to volume with 3 N sulfuric acid. The standard mercury solution, mercury spiking solutions, and the stannous chloride reagent were stored in high-density polyethylene bottles. The 10% potassium permanganate oxidizing solution was prepared by dissolving 100 g potassium permanganate crystals in 1 liter of distilled deionized water.

The column-packing material was Anaport glass beads (80/100 mesh). In earlier studies, the column-packing material employed was Analabs

Chromosorb W-HP (80/100 mesh, 1.5% OV 17, and 1.95% QF 1). Al-though the latter column was equally efficient, it must be conditioned before use at 150°C for approximately 2 hr. No column conditioning, other than normal blank measurements, was necessary with the glass bead column. Colorless silica gel (Fisher Scientific Co., 6/20 mesh) was used as the drying agent.

Procedure

With the U tube column immersed in the liquid nitrogen bath, place a 100 ml sample solution into the gas bubbler. Insert the sparger and switch the valves to the flow through position. The purging rate for the nitrogen aeration is 0.5 liters/min and 7 psi. After purging is completed (7 min), return the valves to the bypass position, and increase the flow rate of the carrier gas to 0.7 liters/min (7 psi). Remove the column from the liquid-nitrogen bath and heat the U tube through the wire windings using the variable transformer. The transformer can be simply switched on by prior calibration of the voltage setting to give a column tempera-ture of 225°C measured on the outside wall of the U tube after 60 sec of heating. The elemental mercury is vaporized and eluted from the col-umn in 1 to 2 sec and the entire operation requires less than 10 sec from the time the U tube is removed from the cold trap. During the heating and elution cycle (60 sec), the flow rate of the carrier gas decreases from 0.7 to 0.15 liters/min. The absorption at the mercury wavelength (2537 Å) occurs in the gas cell 12 sec after the heating cycle is initiated. The absorption is recorded in arbitrary units and the maximum height noted. This represents the sample and system blank.

After the response has returned to the initial base line (60 sec), stop the heating and cool the column in air for 30 sec. Return the U tube to the liquid-nitrogen bath. The carrier gas flow is set to 0.5 liters/min and the system is ready for the stannous chloride reduction and nitrogen aeration step. The total time for this operation is 9 min.

Add 0.5 ml of the stannous chloride reagent to the 100 ml sample to reduce mercury to its elemental state. Mix the sample solution by hand-shaking for 5 sec and then switch the Teflon stopcocks to the flow-through position. After the latter operation, the manipulations are identical to the steps outlined in the procedure for establishing the system blank. Record the absorption peak (2537 Å) of mercury and note its maximum height. Add appropriate spikes of mercuric chloride standards to the sample matrix and repeat the procedure. Three spike additions are usually made. A system blank is repeated and the mercury concentration determined from an individual calibration curve for each sample.

Determination of Metals in Sea Water Using Chelating Resin (73)

Comment on Method

An alternative to solvent extraction preconcentration of metals in waters is ion exchange chromatography. Riley and Taylor point out that anion and cation exchangers are of limited application for the analysis of sea waters. On the other hand, chelating resins show promise for use with a wide range of elements. Recoveries of 99–100% of bismuth, cadmium, cobalt, copper, indium, manganese, molybdenum, nickel, lead, rhenium, scandium, thorium, tungsten, vanadium, yttrium, and zinc are reported by using Chelex 100. The following method was tested on sea water for the atomic absorption analysis of copper, zinc, cadmium, nickel, and cobalt.

Reagents and Equipment

A Techtron AA4 atomic absorption spectrophotometer with digital output was used in the experimental work. Running conditions of the hollow-cathode lamps were those recommended by the manufacturers. The burner was used with coal gas–air and the optimum operating conditions were found empirically for each element. Presumably an air–acetylene flame would also be applicable. In a recent letter the authors of this reference (73) indicate that any good atomic absorption instrument with background correction may be employed.

All glass and silica apparatus should be allowed to stand overnight filled with 1:1 mixture (v/v) of concentrated nitric and sulfuric acids.

Stir Chelex 100 ion exchange resin (50–100 mesh) with excess 2 M nitric acid for 5 min. Wash the resin thoroughly with redistilled water and treat it with excess 2 M ammonia solution (prepared by appropriate dilution of isothermally distilled ammonia). Pack the resin thoroughly using redistilled water and treat it with excess 2 M ammonia solution (prepared by appropriate dilution of isothermally distilled ammonia). Pack the resin into glass columns to give columns 15 cm × 1.5 cm². After use, wash the resin with 80 ml of redistilled water and immediately reconvert it to the ammonium form for reuse. Do not allow the resin to stand in the hydrogen form as it is somewhat unstable.

Distilled water and nitric acid should be redistilled using a silica still. Acetone should be redistilled in an all-glass apparatus.

Dissolve 1.00 g each of copper, nickel, cobalt, zinc, and cadmium in a slight excess of an appropriate acid and dilute each solution to 1 liter. Dilute 10 ml aliquots of these solutions to 100 ml with acetone. The resultant solutions contain 100 μg of the element per milliliter; use these solutions to prepare working standards containing 1.0–5.0 μg of the metals per 10 ml, each solution being 90% with respect to acetone.

Procedure

Filter sea water samples through a 0.5 μm membrane filter and allow 10 liter aliquots to pass through the column of Chelex 100; the flow rate should not exceed 300 ml/hr. If only zinc, copper, and nickel are to be determined, 1 liter aliquots are sufficient. Wash the column with 250 ml of water and reject the washings. Dilute copper, nickel, zinc, and cadmium with 30 ml of 2 N nitric acid and then elute cobalt with 20 ml of 2 N hydrochloric acid. Place the eluate in a 50 ml silica conical flask, cover with silica bubble stoppers, and evaporate to dryness on a hot plate at low temperature. Add to each flask, by means of pipet, 1 ml of 0.1 N nitric acid and when the residue has dissolved add 9 ml of acetone. Determine copper, nickel, cadmium, and cobalt in the appropriate solution using the atomic absorption spectrophotometer and the resonance lines at wavelengths of 3247, 2318, 2287, and 2406 Å, respectively. Dilute 1 ml of acetone solution from the 2 N nitric acid eluate to 5 ml with 9% aqueous acetone and use the resultant solution for the determination of zinc using the 2137 Å zinc resonance line. Determine the reagent blank for the method in the same manner using sea water that has been stripped of trace elements by passage through a column of Chelex 100. Calibrate the instrument in the range of 0.5–2.5 $\mu g/ml$ with standard solutions of the elements in 90% acetone; 90% acetone is used as a blank for the calibration.

Determination of Arsenic, Germanium, Antimony, Selenium, Tin, Bismuth, and Tellurium Using a Conventional Generator (74)

Comment on the Method

The authors of this procedure have carefully studied the optimum reaction times for the various elements. The times given apply to water samples and they caution the worker that longer reaction periods may be necessary if other sample types are analyzed.

The effect of hydride collection time (time period between adding the sodium borohydride and sweeping the generated gases into the flame) was investigated for arsenic, bismuth, germanium, antimony, selenium, tin, and tellurium. Only for the analysis of bismuth and tellurium is there a need for careful monitoring of the collection time. A significant loss of sensitivity is obtained using collection times longer than

30 sec. By keeping the four-way stopcock in the sweep position, one may continuously flush the tellurium hydride into the burner. This degrades the sensitivity by about a factor of 2; however, the need for time monitoring is eliminated.

For other elements determined successfully using the sodium borohydride method (arsenic, germanium, antimony, selenium, and tin), collection times of up to 2–3 min can be used with no loss of sensitivity. Generally, a collection time of 30 sec is sufficient, although longer collection times may be required for certain types of samples. The sodium borohydride pellet reacts vigorously when added to the acidified sample, and dissolves completely in about 20 sec. The addition of a single sodium borohydride pellet (10/32 in.) weighing approximately 200 mg was found to provide optimum results. Varying the amount of sodium borohydride from 100 to 400 mg had no noticeable effect on the sensitivity obtained for any of the elements studied. The addition of amounts larger than 500 mg caused the balloon reservoir to rupture. The use of a magnetic stirrer to agitate a sample solution of arsenic had no effect on the sensitivity obtained.

For arsenic, bismuth, antimony, and tellurium varying the acid concentration from 1 to 6 N had no noticeable effect on the sensitivity obtained. To obtain optimum sensitivity, the acid concentration should be at least 5 N when determining selenium and 4 N or less when determining germanium. For the determination of tin, it is important that the acid concentration be 0.5 N or less, due to the very pronounced loss of sensitivity with increasing acid concentration. The suitability of other acids was not investigated.

For arsenic, the electrodeless-discharge lamp (EDL) provides about a two-fold improvement of sensitivity, while for selenium the sensitivity is improved by about 30%. The sensitivity for germanium is about an order of magnitude poorer than that obtained for the other elements studied. The absolute sensitivities (weight of an element that gives a signal of 1% absorption) obtained for the seven elements studied are summarized in Table XII together with solution detection limits.

Reagents and Equipment

All results were obtained using a Perkin-Elmer Model 403 atomic absorption spectrophotometer equipped with a three-slot burner head, deuterium background corrector, Model 056 recorder, and Intensitron hollow-cathode lamps. For the determination of arsenic and selenium, Perkin-Elmer electrodeless-discharge lamps were also employed. A Perkin-Elmer arsenic/selenium sampling system was utilized for the generation and collection of the gaseous hydrides. This system utilizes

Table XII

Absolute Sensitivities Obtained Utilizing Sodium Borohydride
Reduction

Element	λ (Å)	Spectral slit (Å)	Absolute sensitivity (ng)	Solution detection limit[a] (μg/liter)	Remarks
As	1937	7	10	0.5	Hollow cathode
			5	0.15	EDL
Bi	2231	2	8	0.25	Collect 30 sec
			12	0.4	Continuous flow
Ge	2651	2	270	10.0	
Sb	2176	2	10	0.25	
Se	1960	7	11	0.25	Hollow cathode
			9	0.15	EDL
Sn	2246	2	7	0.2	
Te	2143	2	14	0.25	Collect 30 sec
			27	0.75	Continuous flow

[a] Based on a 20 ml sample.

a dosing stopcock for reagent introduction and a balloon reservoir for
collection of the generated gases. By rotating a four-way stopcock, the
argon flow can be set to bypass or flow through the generation flask (125
ml Erlenmeyer with a 29/42 ground glass joint). The collected hydride,
plus excess hydrogen, is introduced into the burner via the auxiliary
oxidant connection. The following flow settings were found to be optimal: argon, 40 (13 liters/min) at a pressure of 20 psi; hydrogen, 24 (10
liters/min) at a pressure of 20 psi. Reagents used were $NaBH_4$ pellets
(10/32 in., available from Alfa Inorganics) and hydrochloric acid.
Standard solutions of all the elements investigated were prepared in
dilute HCl.

Procedure

Pipet 20 ml of sample into the generation flask. Acidify the sample
using a suitable volume of HCl and dilute to 40 ml with deionized
water. Connect the flask to the generation apparatus and open the
four-way stopcock for about 15 sec to admit argon, which flushes the air
out of the system. After flushing, close the four-way stopcock and add a
single sodium borohydride pellet via the dosing stopcock. Continue the
reaction for a time that will vary depending on the element being
determined and the type of sample being analyzed. For aqueous samples, a reaction time of 30 sec is suitable. Longer reaction times may be
required for some types of samples. Open the four-way stopcock, which

allows the auxiliary argon flow to sweep the generated gases into the burner. Obtain the absorption signal on the recorder. Close the four-way stopcock after the absorption signal has been recorded, and the pen has returned to the baseline. Analyze standards, including a reagent blank, using the same procedure.

When the four-way stopcock is opened, the surge of excess hydrogen into the flame causes a sudden change in the absorption of the flame, which produces a large blank signal when operating at wavelengths below 2100 Å. Use of the deuterium background corrector appreciably reduces this blank signal.

Determination of Metals in Freshwater Using Electrothermal Atomization (75)

Comment on the Method

As already stated, it is not advisable to analyze waters with high salt content (e.g., seawater, brackish water, and some effluents) directly by electrothermal atomization. Problems with nonspecific absorption are normally too great for reliable automatic background correction. The following procedure should be used for samples such as rain, snow, soft-water lakes, rivers, and drinking waters.

It is surprising to find that values obtained by direct electrothermal analysis are sometimes different from those obtained by extraction or ion exchange. The author believes this to be due to the presence of fine particulate even after filtration through 0.45 μm filters.

Loss of metals during ashing can occasionally be a problem. This is often due to the presence of abundant chloride. Nitric acid is therefore added to the samples to convert metals to the oxide during this stage. Drop by drop, addition of 50% hydrogen peroxide can also aid in the ashing and in this type of oxidation and may be used as necessary.

Reagents and Equipment

The following procedure is strictly for use with the HGA 2100 (HGA 74) furnace from Perkin-Elmer. Other Perkin-Elmer equipment may require different conditions. It is almost certain that other manufacturers' furnaces, rods, etc., will require different conditions. These can usually be found experimentally starting with those recommended by the manufacturer for standard solutions. Both Perkin-Elmer 305B and 603 atomic absorption units have been used. A Perkin-Elmer Model 056 recorder was employed. Background correction is essential.

Samples were injected using disposable Eppendorf pipets. The disposable plastic tips were treated in 3 M nitric acid and rinsed with water to eliminate contamination of iron and zinc. The largest single volume used was 50 μl.

The quality of the distilled water is crucial in this work. Detection limits are often set too high because of impurities in distilled water. Double distillation in quartz will normally give a suitable high-quality water.

Stock standard solutions (1000 μg/ml) are prepared by dissolving pure metals or pure metal salts in a minimum of acid and then diluting to volume in distilled water. Nitric acid should always be used. Dilute working solutions of mixed cadmium, cobalt, chromium, copper, iron, manganese, nickel, lead, and zinc are prepared in concentrations from 0.3 to 100 μg/liter and are made to contain 1% nitric acid. Contamination at lower concentrations necessitates the renewal of standards at frequent intervals.

Procedure

Using an appropriate-sized pipet, place standards and then samples into the furnace one after the other. All waters should contain or be treated with nitric acid to prevent formation of volatile metal salts during ashing.

The atomic absorption equipment including the background corrector should be warmed up for 20 min or until sources are stabilized. Use temperature programs and instrument parameters as outlined in Table XIII. Use nitrogen purge gas in the normal flow mode.

Run one or two standards every 10–20 samples to check the stability of the readings obtained. Recalibrate if necessary. When going from high to low values, rinse the pipet. It is a good precaution to rinse between each sample to prevent spurious contamination.

Table XIII

Temperature Program—AAS Parameters, General Conditions

Element	Wavelength (Å)	Slit (Å)	Dry (°C)	Ash (°C)	Atomize (°C)
Cd	2288	7	100	440	1800
Co	2407	2	100	1000	2600
Cr	3579	7	100	1000	2700
Cu	3247	7	100	1000	2500
Fe	2483	2	100	1200	2600
Mn	2795	2	100	1000	2500
Ni	2320	2	100	1200	2600
Pb	2833	7	100	500	2000
Zn	2138	7	100	500	2000

3
Analysis of Geological Materials

A study by Fairburn (76) in 1951 cast serious doubt on the reliability of geological chemical data. Anyone contemplating the analysis of geological samples should read this classic paper. Although pertaining to silicate rocks, the study created a good deal of skepticism about analytical chemical data in general. Following the publication of this important study, analytical chemists have spent a much greater proportion of research effort on improving the precision and accuracy of chemical analyses.

Atomic absorption has been applied to the analysis of both major and minor elements in geological samples. The technique is more useful for the latter. Precision and accuracy requirements of major-element analysis of rocks and minerals are often so stringent that they are extremely difficult to achieve by atomic absorption spectroscopy. Despite this fact, many laboratories, out of necessity, employ atomic absorption for whole-rock analysis.

It is a waste of time to analyze samples that have been improperly taken and/or improperly physically prepared. Sampling and physical preparation are beyond the scope of this book, but in the author's opinion, improper sampling and physical preparation are very common problems in the geological sciences.

SAMPLE PREPARATION

Drying, Grinding, and Sieving

In most cases, samples must be taken into solution prior to atomic absorption analysis. To prepare material for decomposition, drying, grinding, and sieving must be done. For most purposes, samples are first dried at 100°C in air. Rocks, ores, and minerals are then ground to a grain size that will pass a 200 mesh sieve. There is no general rule for the sieving of soils. Geochemical exploration and environmental applications usually require an 80 mesh fraction, but frequently 200 mesh and even finer size fractions are required. When mercury is to be determined it is important to avoid the high temperatures often encountered in conventional drying and grinding procedures. In general, temperatures above 80°C are to be avoided for mercury.

Sample Decomposition

It is important in most cases to obtain the material completely in solution prior to atomic absorption analysis. This is almost always true for rocks, minerals, ores, meteorites, concentrates, and organic samples. To accomplish this, either a fusion or acid decomposition may be used. With atomic absorption, particularly when electrothermal methods are to be used, it is important to keep the total salt content as low as possible. For this reason acid decomposition is commonly used.

When silicate matter is to be analyzed and when silicon is not of interest, a mixture of hydrofluoric and other mineral acids can be employed in an open dish. If silicon must be determined, a hydrofluoric acid–mineral acid mixture in a closed vessel, or a fusion is suitable. In the case of the latter, lithium metaborate or lithium tetraborate fusion followed by addition of hydrofluoric and boric acids to stabilize silicon has been utilized. A sodium peroxide sinter will decompose almost all silicates and may be recommended. Other fusing agents either do not attack a broad range of samples or are unacceptable for other reasons, e.g., excessively high fusion temperature or purity of reagent.

For sulfide matter, acid attack using a combination of mineral acids is preferred. An oxidizing fusion using sodium peroxide is often very effective but leads to high salt contents. It may be important to dissolve only sulfide material in the presence of other metal-bearing minerals. For this purpose an attack using a buffered solution of hydrogen peroxide may be employed. The specificity of a sulfide metal decomposition is open to question and the author advises caution. Oxide minerals with the rutile structure, e.g., chromite, are particularly resistant to attack. The pressurized bomb technique using a combination of acids can be recommended.

Removal of Organic Matter

Samples high in organic matter must be specially treated. The destruction of organic matter is accomplished by either a wet or dry ashing technique. Dry ashing can be done when mercury, arsenic, or selenium is not of interest. Losses of other elements such as cadmium and zinc have been confirmed above 450°C. Low-temperature plasma ashing devices for dry ashing exist. This equipment is costly and ashing usually requires long time periods.

Wet ashing is the procedure preferred by most workers for treatment of organic samples. Treatment with a perchloric–nitric acid mixture is usually employed. To minimize the possibility of an explosion when using perchloric acid, an excess of nitric acid must be present at the outset. Perchlorates are undesirable in solutions used for electrothermal atomic absorption. It is advisable, in this case, to fume off excess perchloric acid. Organic material can contain appreciable silicate. Hence, for a total decomposition it may be necessary to add hydrofluoric acid to the perchloric acid mixture.

Special Considerations for Soil Analysis

In soil analysis it is seldom desirable to obtain a total decomposition. For both exploration geochemistry and environmental analyses the metal in the clay fraction is of interest. A mixture of mineral acids, commonly hydrochloric and nitric acids, is employed to dissolve this metal.

Recently it has become important to measure the "available" metal in soils. This means the fraction that would be available to plants when plant uptake is of interest. Some of the extractants proposed for this purpose are acetic acid, ethylene diamine tetraacetic acid (EDTA), nitrilotriacetic acid (NTA), and citric acid. Research at present suggests that the best correlation between the metal uptake and a reagent depends on the particular metal and its chemical combination in the soil. A great deal more research is essential to allow a quantitative explanation of "available" metals. In this regard, approaches outlined in the chapter on metal speciation may be of use.

Elements Requiring Special Consideration during Sample Decomposition

Decomposition of samples when mercury and/or chalconides (arsenic, selenium, tellurium, antimony, and bismuth) are to be analyzed requires special conditions. In the case of mercury, oxidizing conditions throughout the decomposition are ensured usually by the addition of potassium permanganate or potassium persulfate. For the chalconides an oxidizing acid decomposition employing potassium persulfate or a fusion with potassium hydroxide is satisfactory.

The analytical chemistry of the platinum metals is complex. This statement remains true in atomic absorption analysis. Atomic absorption sensitivities are such that direct analyses of acid extracts from rocks and ores are not possible. For these reasons, and due to inhomogeneity problems, it is essential that a fire assay pretreatment of ores and rocks be performed prior to atomic absorption analysis.

Special Precautions

Contamination can be a serious problem during sample preparation. It is very important to run reagent blanks side by side with the samples throughout all procedural steps. The chemistry of the elements to be analyzed must always be considered before adopting a sample decomposition procedure. For example, sulfuric and hydrochloric acids should not be used when lead and silver, respectively, in other than sub submicrogram/gram amounts are to be determined. Reagents containing chloride should also be avoided when solutions are to be analyzed by electrothermal atomization.

DETERMINATION OF MAJOR AND MINOR ELEMENTS IN ROCKS

Analysis of Minerals and Rocks—Sodium Peroxide Sinter (77)

Comment on the Method

The authors of this procedure found that other decomposition methods failed to solubilize a number of important sample types. The peroxide sinter method given below has been shown to decompose a wide variety of difficult-to-dissolve geological materials. Trouble was encountered only in the case of a magnesite-chromite. The authors stress the need for strict attention to several details of the technique. This method is selected for detailed treatment here because it covers the widest range of sample types and elements.

Reagents and Equipment

The sodium peroxide must be fresh, dry, and in powder form; the Merck reagent has been found suitable. A fourfold excess over the weight of sample is used. The sample must be dry and ground to 300 mesh. Tektron atomic absorption equipment was used.

Procedure

An intimate mix of sample and sodium peroxide must be effected quickly before the sodium peroxide becomes moist. This can be achieved in several ways. One method is to transfer the weighed material quickly to a dry small glass tube with a ground glass stopper and mix by shaking. Another method is to weigh the sample on a glass weighing vessel, weigh the peroxide directly into the crucible using a top-loading balance, and quickly transfer the sample onto the top of the peroxide and mix rapidly with a plastic rod.

Platinum or platinum–gold crucibles are used when the sulfur content is below 5%. Nickel, iron, or zirconium crucibles can be used irrespective of the sulfur content, depending upon the elements to be determined. There is no evidence that platinum could not be used for higher sulfur contents, when lead, copper, or nickel are not present. The avoidance of platinum crucibles is merely a precautionary measure. Platinum and platinum–gold crucibles have been used constantly for 2 years without serious weight loss.

A temperature of 480°C will decompose most minerals. Many minerals will decompose at 430°C or less. At 480°C, a treatment time of 2 hr will decompose the most resistant minerals, although 20 min is sufficient in most cases. At lower temperatures longer times may be necessary.

Care has to be taken in the method of dissolution to prevent hydrolysis of silica. Extract the sintered material in cold water and add water to bring the volume of the solution to slightly below the final volume required. Add 5 M hydrochloric acid until the solution clears, and add excess hydrochloric acid to make the solution 0.3–0.4 M. Samples high in manganese can be brought completely into solution by the addition of two or three drops of hydrogen peroxide to the acidified solution.

See Tables XIV and XV for the elements determined using nitrous oxide–acetylene and air–acetylene gas mixtures, respectively. The conditions used and the additions made to overcome interferences up to the level shown are also given. The high concentration of sodium acts as an ionization buffer, but for best results the additions listed are necessary.

The interference suppressors are added to the sample aliquot prepared for spraying, and to the calibration solutions. In all cases the same aliquot and the calibration solutions contain the same amount of sodium, all dilutions being with a 0.3–0.4 M hydrochloric acid solution containing 10 g/liter sodium peroxide. The solution must be boiled to remove free oxygen and prevent bubbles forming during nebulization.

The addition of lanthanum for the determination of aluminum is not necessary to overcome interference, but its presence improves flame stability, thus permitting a wide choice of flame conditions.

The use of an air–acetylene flame for the determination of calcium

Table XIV

Nitrous Oxide–Acetylene Flame[a]

Element	Burner height below light path (mm)	Height of red zone (mm)	Typical concentration (μg/ml), absorbance = 0.4	Maximum concentration tested for interferences (μg/ml)	Amount of interference suppressor added
Si	2	12	160	Fe 600, Al 250, Ca 200, Mg 200, Cr 450	1% La
Al	5	12	110	Fe 600, Ca 200, Mg 200, Cr 450, Si 400	Not necessary[a]
Fe	8	5	180[b]	Al 500, Si 400, Ca 200, Mg 200, Cr 200	1% tartaric acid
Cr	2	5	16	Fe 600, Si 400, Ca 200, Mg 200, Al 500	1% tartaric acid
Ti	2	5	200	Fe 600, Si 400, Ca 200, Mg 200, Al 500, Cr 450	0.02% Al

[a] The addition of 1% La improves flame stability.
[b] Using line 3719.9 Å.

Table XV

Air–Acetylene Flame[a]

Element	Burner height below light path (mm)	Typical concentration (μg/ml), absorbance = 0.4	Maximum concentration tested for interferences (μg/ml)	Amount of interference suppressor added
Ca	2	7	Fe 200, Al 500, Si 400, Mg 200, Cr 450	1% La
Mg	2	1.8	Fe 200, Al 500, Si 400, Ca 200, Cr 450	1% La
Cu	2	5	Fe 600, Al 500, Si 400, Ca 200, Hg 200, Cr 450	Not necessary
Pb	2	40[b]	As for Cu	Not necessary
Zn	2	1	As for Cu[c]	Not necessary
Co	2	8	As for Cu	Not necessary
Ni	2	8	As for Cu	Not necessary
Mn	2	3	As for Cu	Not necessary

[a] All solutions contain 7500 μg/ml sodium.

[b] Using line 2833 Å.

[c] Where iron is high, as in iron ores, iron can interfere, and a background correction is necessary.

and magnesium is preferred to the high-temperature nitrous oxide–acetylene flame because greater stability is achieved despite a loss of sensitivity. In the literature there has been an unwarranted extrapolation of the original comments of Amos and Willis (78) with respect to the removal of interferences in the determination of calcium and magnesium in the high-temperature flame. They did not claim that all interferences are removed, but rather that they are much reduced. It has been found necessary to add lanthanum when using the high-temperature flame to determine calcium and magnesium under these conditions.

Analysis of Minerals and Rocks—Lithium Fluoroborate Method (79)

Comment on the Method

This procedure is an alternative to the sodium peroxide sinter method given above. The author's experience suggests that the method can be highly recommended.

Calibration is accomplished through the use of solutions of standard reference rocks and minerals. Standards used were from the United States Geological Survey, U.S. Department of the Interior, Reston, Virginia 22092; Bureau of Analysed Samples, Newham Hall, Newby, Middlesbrough, Teeside TS8 9EA, England; National Bureau of Standards, Office of Standard Reference Materials, Room B311, Chemistry Building, National Bureau of Standards, Gaithersburg, Maryland 20234; and Center de Researches Petrographiques et Geochimiques, Case Officielle N° 1, 54500 Vandoeuvre-les Nancy, France. Advantages in the use of standard solutions of this type are a reduction in the total number of standard solutions required and the knowns and unknowns are of similar composition and undergo identical chemical treatment.

The presence of fluoroborate suppresses some of the chemical interferences commonly found in atomic absorption work (e.g., the interference of aluminum and silicon in the determination of barium and strontium). This is presumably due to the complexation of the interferents as fluoroanions. Another advantage found for the presence of fluoroborate was the avoidance of problems due to titanium hydrolysis. As a result, the high acidities usually required to stabilize titanium solutions were avoided.

Atomic absorption measurements are done as follows: after the first approximation of the concentration is found by comparison to any convenient standard, the samples are arranged in descending order of apparent concentration of the desired element. A group of at least four standards is selected, such that the concentration of the desired element in the lowest standard is lower than that in the sample with the lowest concentration, and similarly for the highest standard. Atomic absorption readings are then taken for both standards and samples, grouped together in descending order of concentration, and repeated in ascending order. Where readings are not sufficiently consistent, the entire cycle is repeated. For each set of readings in either direction, a curve of absorbance vs. concentration is plotted from the readings on the standards (or, where possible, the best straight line is calculated), and results for the samples derived from the calibration *for that set of readings.* Final results are based on averaging two or more calculated values for each element in each sample. Advantages of this measuring technique are that errors in the standards are smoothed out in plotting the curve, and instrumental drift is essentially compensated for by measuring samples and standards at the same time.

A Goquel burner (80) was used throughout this work for both air and nitrous oxide–acetylene flames. Water cooling is a feature of this burner, with the result that the burner is never too hot to touch. The slot length is only 7.4 cm but sensitivities with air–acetylene are comparable to those obtained with conventional burners having a 10 cm slot. Using nitrous oxide–acetylene, the sensitivities are equal to or greater than conventional nitrous oxide, 5–6 cm slot burners.

Reagents and Equipment

The atomic absorption instrument used was a modernized Techtron Model AA-3 equipped with a homemade Goquel burner (80). The type of hollow-cathode lamp used is listed in Table XVI.

Fusions were made in graphite high-purity graphite crucibles. These are available from Ultra Carbon, Bay City, Michigan, under catalogue number C-682-1B. The cover for the crucible is catalogue number A-6206.

Hydrofluoric acid (6:19). Mix 120 ml of concentrated hydrofluoric acid with 380 ml of water. Store in a plastic bottle.

Boric acid, 50 g/liter. Dissolve 50 g of boric acid crystals in about 400 ml of boiling water in a silica or Vycor beaker, with magnetic stirring. Pour this into about 500 ml of cold water in a 1000 ml polyethylene graduate cylinder. Stir to mix and dilute to volume. Store in a plastic bottle. Because this solution is used in large quantities, several liters should be prepared at the same time.

Table XVI

Operating Parameters

Component	In 100 ml mg sample	mg Sr	Range (percent in sample)	HC lamp Type	mA	Support gas	Band-pass (Å)	Wave-length (Å)	Notes
Al	50	300	0–25	Si–Al	15	N₂O	3.3	3093	Total iron, expressed as Fe₂O₃
Fe	10	150	1–30	Fe	5	Air	1.7	2483	
Mg	10	150	0–15	Ca–Mg	5	Air	3.3	2852	
	10	150	5–50	Ca–Mg	5	N₂O	3.3	2852	
Ca	10	150	0–3	Ca–Mg	5	N₂O	3.3	4227	
	10	150	2–20	Ca–Mg	5	Air	3.3	4227	
Na	10	150	0–10	Na–K	10	Air	1.7	5890	
K	10	150	0–10	Na–K	10	Air	3.3	7665	
Ti	Note	0	0.05–10.0	Ti	20	N₂O	1.7	3643	10 mg sample, 10 mg Al in 11 ml
Mn	50	300	0.01–1.0	Mn	5	Air	1.7	2795	
Ba	Note	0	0.01–1.0	Ba	10	N₂O	3.3	5536	⎫ 10 mg sample, 50 mg NaCl in 10.5 ml
Sr	Note	0	0.01–1.0	Sr	10	N₂O	3.3	4607	⎬
Cr	50	300	0.01–1.0	Cr–Ni	8	N₂O	3.3	3579	⎭
Ni	50	300	0.01–1.0	Ni	8	Air	1.7	2320	

Strontium nitrate, 15,000 μg Sr/ml.   Dissolve 72 g of strontium nitrate in about 1 liter of water and dilute to 2 liters. Store in a plastic bottle.

Aluminum buffer.   Dissolve 14 g of Al $(NO_3)_3$ $9H_2O$ in a minimum volume of water (25–30 ml) in a 400 ml beaker. Add 100 ml concentrated hydrochloric acid, cover, and boil until no further gases are evolved. Evaporate to dryness overnight on a steam bath. Add 25 ml of concentrated hydrochloric acid and just enough water to provide a clear solution. Repeat the boiling and evaporation to dryness. Add 10 ml of concentrated hydrochloric acid, about 50 ml of water and warm to dissolve. Dilute to 100 ml.

Sodium buffer.   Dissolve 10 g of sodium chloride in water and dilute to 100 ml.

Blank solutions.   Stir 1.0 g of lithium metaborate in 50 ml of water in a 400 ml plastic beaker, and add 25 ml of hydrofluoric acid (6:19). Add 100 ml of boric acid solution (50 g/liter), stir to complete solution, and dilute to 200 ml. Store in a plastic bottle. For each size of aliquot of sample solution required for analysis, a corresponding blank solution may be prepared from the master blank solution, including the same additives. Large quantities of the master blank solution are required in preparing standard silica solutions, so it may be necessary to prepare a larger volume of the former.

Procedure

Sample solutions.   All "standard" samples and other samples that are to be analyzed on the "dry basis" should be weighed into small porcelain crucibles and dried at 110°C. The sample weight (after drying, where that is done) should be between 200 and 205 mg, weighed to the nearest 0.1 mg. For some samples, decomposition will be facilitated if the sample is preignited at 600–700°C, preferably overnight. Samples may be conveniently analyzed in groups of six. Preignite the covered graphite crucibles at about 1000°C for 15–20 min and cool. Without disturbing the graphite dust in the crucible, add 1.0 g of lithium metaborate. Carefully brush the weighed sample onto the lithium metaborate and mix lightly with a small spatula. Place the covered crucible in a preheated muffle furnace and hold at 950–1000°C for 15 min. Quickly but carefully move the crucible to the front of the furnace, remove the lid, and take the crucible out of the furnace. Immediately rotate the molten bead to pick up any unattacked sample and allow to cool until the crucible no longer glows. Return the covered crucible to the furnace (reversing positions where a group is involved) and heat for 5 min more.

While the fusion is going, remove the screw-cap from a 250 ml transparent plastic (e.g., trimethylpentene) jar and place 40 ml of water in the jar. Remove the crucible from the furnace as done before, but this time pour the hot, molten fusion into the water in the plastic jar. After the crucible has cooled to room temperature, examine it carefully for any retained material from the fusion. If any is found, use a small spatula and a brush to transfer it to the plastic jar.

Add a Teflon-coated stirring bar to the jar, followed by 25 ml of hydrofluoric acid, and immediately cap the jar tightly. Stir the contents until the shattered fusion bead is completely disintegrated, and then chill thoroughly in a refrigerator.

Open the chilled jar, immediately add 100 ml of boric acid solution (50 g/liter), cover again, and stir until the solution clears (except for a few black graphite particles from the crucible). Filter through a rapid filter paper (e.g., Whatman No. 541) in a plastic funnel, receiving the filtrate in a 200 ml volumetric flask. Wash the jar and the filter several times with water.

Dilute the filtrate to the mark, add m ml of water, where $(200 + m)$ mg is the sample weight. Mix well and transfer to a polyethylene bottle. (Teflon bottles are recommended for the "standard samples.") The final solution then contains 200 mg of sample per 200 ml of solution, and the bottle should be labeled with the sample number and date.

Atomic absorption measurements.   After sufficient warm-up, adjust the parameters for the determination of each element as listed in Table XVI. Aspirate a standard solution and optimize other parameters (gas controls, burner position, damping, scale expansion) for each element.

For each element, obtain a first approximation of its concentration by comparison with any convenient standard. Select at least four standards, covering a concentration range slightly greater than that of the samples. Arrange samples and standards together in a series of progressively decreasing concentration of the desired element. Measure the absorbance of all of the solutions in order (preferably with a recorder), then repeat the readings in the reverse order. For elements whose oxides are present to the extent of 10% or more, the entire sequence of measurements should be repeated. This should also be done where there are apparent discrepancies between the two sets of readings.

For each set of readings, plot a curve (or calculate the best straight line) from the absorbances and known concentrations of the standards, and read values for the samples from the curve. Calculate the mean of all values for a given element in a given sample, rejecting any questionable readings. If there is any doubt about the acceptability of a particular reading, additional measurements should be made.

Decomposition Method for Soils and Rocks Containing Chromite

Methods given above may not decompose minerals with the spinel-type structure. If this problem is encountered a fusion with sodium pyrosulfate should be used (81).

DETERMINATION OF MINOR ELEMENTS IN GEOLOGICAL SAMPLES—TOTAL RECOVERY

Determination of Copper, Zinc, and Lead in Geological Materials (82)

Comment on the Method

Copper and zinc are analyzed directly in the sample solution. Lead is extracted into trioctylphosphine oxide–methyl isobutyl ketone prior to analysis. A hydrofluoric acid–hydrochloric acid–hydrogen peroxide sample digestion is used to ensure freedom from residual nitric acid, which causes errors in the TOPO–MIBK extraction. Had this problem not been present, hydrofluoric–nitric acid digestion would have sufficed.

The procedure recommends turning the burner head to reduce sensitivity. The author finds that this practice can sometimes lead to unacceptably noisy results. Other alternatives would be to dilute the samples or use a less-sensitive line. The procedure was tested using spiked samples, which is not as satisfactory as using standard reference materials.

Reagents and Equipment

A Varian AA6 atomic absorption spectrophotometer was used for all determinations with the instrument settings shown in Table XVII. The air pressure was 45 psi (meter reading 4.0) and the acetylene pressure 10 psi (meter reading 1.2 for copper and zinc and 0.2 for lead) in all cases. The AA6 was equipped with a simultaneous background corrector, an automatic gas control unit, a variable nebulizer, and Westinghouse

Table XVII

Instrument Settings

Metal	Lamp current (mA)	Wavelength (Å)	Slit (Å)	Burner height (div)
Cu	2.5	3247	5	12.0
Zn	2.0	2139	5	14.0
Pb	5.0	2170	10	11.5

hollow-cathode lamps. All measurements were made in concentration, integration modes, and using the curve corrector. The aspiration rate and glass bead position were adjusted for maximum absorbance and minimum noise. The burner head was adjusted to an angle of approximately 20° from the parallel position for copper, and to an angle of 90° for zinc readings. These burner positions provided a working range of 1–20 μg/ml for the two metals. The position of the burner was parallel to the light beam for lead determinations, given a working range of 1–10 μg/ml.

Iodide reagent. Prepare daily a solution containing 30% (w/v) potassium iodide and 10% (w/v) ascorbic acid in 10% (v/v) hydrochloric acid.

Trioctylphosphine oxide—methyl isobutyl ketone solution. Dissolve 5 g of trioctylphosphine oxide in 100 ml of methylisobutyl ketone.

Copper, zinc, and lead stock solutions. Prepare individual 1000 μg/ml stock solutions by dissolving appropriate amounts of Specpure metal oxides (1.252 g CuO, 1.245 g ZnO, or 1.077 g PbO) in 100 ml of concentrated hydrochloric acid and dilute this to 1 liter with water. A tenfold dilution of the stock solutions with 10% hydrochloric acid gives 100 μg/ml solutions of the respective metal. For combined copper and zinc standard solutions (0, 1, 5, 10, and 20 μg/ml for each metal), pipet 0, 1, 5, 10, and 20 ml of the above 100 μg/ml solutions into five 10 ml volumetric flasks and dilute to volume with 10% hydrochloric acid.

Lead standard solutions in methylisobutyl ketone (0, 1, 5, and 10 μg/ml Pb). First, prepare an aqueous lead standard solution (10 μg/ml) by dilution with 10% hydrochloric acid. Pipet 0, 1, 5, and 10 ml of 10 μg Pb/ml solution into four 25 × 150 mm screw-cap tubes along with 10 ml of iodide reagent and 2 g of ascorbic acid, and dilute to 30 ml with 10% hydrochloric acid. Shake to dissolve the ascorbic acid, add 10 ml of trioctylphosphine oxide–methyl isobutyl ketone solution, and shake for 30 sec. The organic lead standard solutions are stable for at least 48 hr.

Procedure

Weigh 0.50 g of less than 100 mesh soil, rock, or steam-sediment sample into a 100 ml Teflon beaker. Wet the sample with 2 ml of water and add 5 ml of concentrated hydrochloric acid and 10 ml of 50% hydrofluoric acid. Place the beaker on a hot plate preset at 130°C and evaporate the solution to incipient dryness. Remove the beaker from the hot plate and allow it to cool. If the solution is not clear after the above hydrofluoric–hydrochloric acid treatment, add 5 ml of concentrated hydrochloric acid and 5 ml of 30% hydrogen peroxide to the residue and allow about 10 min for the reaction to subside. Evaporate the solution to dryness. Add 5 ml of concentrated hydrochloric acid to the cool residue, loosen the residue with a Teflon spatula to aid in dissolution, and return the beaker to the hot plate for 3 to 4 min with occasional swirling. Transfer the solution and any remaining residue to a 25 × 200 mm screw-cap tube marked for a 50 ml volume. Rinse the beaker with water, add the rinse water to the tube, and dilute to the 50 ml mark with water. Centrifuge the tube until a clear supernatant solution is obtained. Alternatively the suspended sediment can be filtered or allowed to settle overnight. Aspirate the solution into the air–acetylene flame of the spectrophotometer for determinations of copper and zinc.

For the determination of lead, transfer one 20 ml aliquot of the clear sample solution to the marked 25 × 100 mm screw-cap tube. Add 2 g of ascorbic acid and 10 ml of iodide reagent. Shake to dissolve the ascorbic acid. Add 10 ml of trioctylphosphine oxide–methyl isobutyl ketone solution and shake for 30 sec. Allow the organic and aqueous phases to separate, and aspirate the upper organic layer into the flame.

Determination of Tellurium in Silicate Rocks (83)

Comment on the Method

Hydride generation is accomplished in a simple apparatus consisting of a 25 ml generator flask and 50 ml test tube trap. Atomization occurs in a heated quartz cell. The procedure has been designed to allow variation in experimental parameters without adverse effect. For example, sample volume and borohydride concentration can vary widely from one run to the next. Increased sensitivity is obtainable by eliminating the spray trap and altering the nitrogen flow. However, these changes require a very careful attention to procedural details and are not adopted.

The interference due iron and copper was examined. Up to 100 mg of iron as the iron III chloride could be tolerated without significant problems. Suppression of tellurium absorption occurred in the case of copper, added as the copper II chloride, at 50 μg.

The method was used on several USGS rocks. A comparison of results obtained by this procedure and with other procedures was possible for GSP-1 and BCR-1. Agreement was satisfactory.

Reagents and Equipment

The apparatus (Fig. 28) consists of a 25-ml generating flask (2) with a side-arm (3), through which the borohydride solution is injected from an automatic pipet. The generated hydrides are swept in the nitrogen stream (4–4.5 liters/min) through a 50 ml test tube (4), serving as a spray trap, to a resistance heated quartz cell (6) mounted in an atomic absorption spectrometer. The quartz spectrophotometer cell is 2 cm in diameter with a 150 cm light path; it is wrapped (5) with 52 turns of No. 20 kanthal resistance wire. The cell temperature is controlled with a variac adjusted slightly higher than required for the maximum tellurium sensitivity; this temperature, estimated from the color of the cell, is about 1000°C. The interunit connections are made with plastic tubing and are kept as short as possible. Operating conditions of the atomic absorption spectrometer are those recommended by the manufacturer, and readout is by means of a recorder.

Dissolve tellurium dioxide in a minimum of nitric and hydrochloric acids and then dilute with 6 M hydrochloric acid to give a stock solution of about 100 μg Te/ml. Just before use, prepare a 100 μg/liter solution by dilution of an aliquot of the stock with 6 M hydrochloric acid.

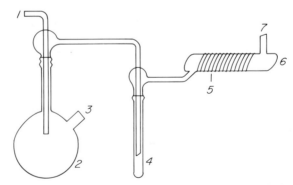

Fig. 28.   Apparatus for the determination of tellurium (83).

Procedure

Weigh a 0.25 g sample into a Teflon beaker, add 5 ml of hydrofluoric acid, and evaporate the contents to dryness at about 180°C. Add 15 ml of hydrofluoric acid, 5 ml of nitric acid, and 0.5 ml of perchloric acid to the beaker, and evaporate to dryness at about 120°C. Add 5 ml of 6 M hydrochloric acid and heat the beaker, covered with a watch glass, at about 80°C for 30–60 min to dissolve the residue. (Rock solutions prepared this way are quite viscous but essentially clear.) Decant the entire rock solution to the hydride-generating flask, and seal the side neck of the flask by insertion of an automatic pipet containing 1 ml of the borohydride solution. When the recorder has returned to a stable baseline, inject the borohydride and measure the peak height of the transient tellurium absorbance signal from the recorder trace.

Then disconnect the generating flask from the system at the ground-glass joint and rinse with water before adding the next sample.

Standards, prepared by adding 0.02–0.1 ml aliquots of the dilute standard solution to 5 ml of 6 M hydrochloric acid in the generating flask are mixed at random through the samples. Sample concentrations are obtained by direct comparison with the peak heights of the standards.

Determination of Minor and Trace Metals in Phosphate Rock Concentrates (84)

Comment on the Method

A combination of atomization techniques are used to cover the 13 elements. Silver, lead, thallium, mercury, and zinc were determined from the solid directly placed in a graphite furnace using solid samples for standardization. Bismuth, cadmium, and indium were determined by the solid sample technique using standard addition of standard solutions. Nickel, chromium, lead, and silver were analyzed in a Varian-Techtron furnace using a standard additions technique. Cadmium, chromium, cobalt, copper, manganese, and zinc were determined in an air–acetylene flame using a standard addition small-volume approach. The procedure was applied to the analysis of phosphate rock concentrates from Florida, Kola (USSR), and Morocco.

Reagents and Equipment

The measurements were made with a Perkin-Elmer 303 or 400S atomic absorption spectrometer. Both instruments were equipped with arc source deuterium lamps for background correction; the 400S instrument also had a tungsten lamp for corrections in the 300–700 nm range.

Solid samples were atomized directly in a homemade graphite furnace heated by a high-frequency generator, the construction of which has been described elsewhere (85). Also see p. 261 for construction details. The high-frequency furnace was used in combination with the Perkin-Elmer 303 instrument. Analyses were based on measuring peak areas.

Sample solutions were atomized in the tube version of the Varian-Techtron Model 63 graphite furnace heated by a homemade power supply. Analyses were done with the Perkin-Elmer 400S instrument and were based on measuring peak heights. The Perkin-Elmer 400S instrument (equipped with a standard single-slot burner) was employed for atomizing sample solutions in an air–acetylene flame. Mercury was determined with the equipment by the procedure described by Omang and Paus (86). Samples were decomposed in polytetrafluoroethylene (PTFE) line aluminum bombs heated on thermostatically controlled hot plates. Samples were weighed on semimicro- or microbalances. All grinding was done in agate mortars. Liquids were introduced into the furnace with 1 and 2 μl ultramico pipets (Oxford Laboratories Inc.) or 10 and 25 μl Hamilton syringes.

Standard solutions were prepared from high-purity metals. Nitric acid was added to dilute standard solutions to maintain pH $\leq$2.0.

All acids were of "Suprapur" quality from Merck.

To prolong the life of the homemade graphite tubes, the solid samples and some of the solid standards were mixed with graphite of the quality used for making the tubes. After being pulverized, the graphite powder was heated for 60 sec at 1950°C.

The solid standards were hydroxyapatite (lead content 2.4 ppm), the U.S. Geological Survey reference rock GSP-1 (thallium content, 1.3 ppm), the Nordic reference rock ASK-Schist (silver content 0.43 ppm), and calcined animal bone (zinc content 193 ppm).

A solid synthetic standard for mercury was prepared by homogenizing known amounts of mercury(II) sulfide and iron(II) sulfide in an agate mortar, the concentration of mercury corresponding to 1.0 μg/g. The content of mercury and the homogeneity of the preparation were checked by analysis.

The particle size of the samples from Kola and Morocco was sufficiently small for the decomposition of about 1 g portions in the bombs, but the sample from Florida had to be ground to pass a 100 mesh sieve.

The very small samples taken for direct atomizations in the graphite furnace made a further reduction of the particle size necessary. Suitable portions of the three samples and three of the solid standards (hy-droxyapatite, GSP-1, and calcined animal bone) were first ground to pass a 270 mesh sieve; 300 mg portions were then intimately mixed in an agate mortar with an equal amount of graphite powder. The solid standard ASK-Schist was ground as already described; however, its content of carbon made it unnecessary to add graphite.

Procedure

Transfer the samples (0.8–1.0 g) to the PTFE vessels of the bombs. Moisten the samples with a few drops of water. Add 1.0 ml of concentrated hydrochloric and 7.0 ml of concentrated nitric acid. After the initial reaction with the carbonates present, close the bombs, heat, and maintain at 180°C for 30 min. After cooling to room temperature, open the bombs, wash the covers and the interior sides with water, and transfer the contents to 50 ml volumetric flasks. Make up to volume with

Table XVIII

Operating Conditions for the Spectrometers and Graphite Furnace

| Element | Wavelength (Å) | Temperatures (°C) and times (sec) of atomization[a] | |
		CRA 63	High-frequency furnace
Ag	3281	2200-5	1900-30
Bi	2231	—	1700-30
Cd	2288	—	1700-30
Co	2407	2600-5	—
Cr	3579	2500-6	—
Cu	3247[b]	—	—
Hg	2537	—	1500-40
In	3039	—	1900-30
Mn	2795[b]	—	—
Ni	2320	2600-5	—
Pb	2833	2000-5	1800-30
Tl	2768	—	1700-30
Zn	2139[b]	—	—
Zn	3076[c]	—	1700-30

[a] For all elements except mercury, and for both furnaces, the atomization step was preceded by 15 sec drying at 80°C.

[b] The wavelength for atomization in the flame.

[c] The wavelength used during measurements with the high-frequency furnace. When this line is employed, a multielement hollow-cathode lamp containing iron should not be used, because it is then difficult to distinguish the zinc line from the adjacent iron line(s).

water. The solutions are stored in plastic bottles. Any residues can be ignored.

Before the start of the measurements, warm up the radiation sources (the deuterium, tungsten, hollow-cathode, and/or electrodeless lamps) for about 0.5 hr. Adjust the flow of argon to 350 and 4.0 ml/min for the homemade and the Varian-Techtron furnaces, respectively. Weigh the solid sample–graphite, or standard–graphite mixture (1–40 mg) in a small tantalum scoop and place in the middle of the homemade furnace by means of a specially constructed adjustable inserting device. Reweigh the scoop and move the furnace to its preadjusted position.

Determine silver, lead, thallium, mercury, and zinc with the high-frequency graphite furnace, using the solid standards. Determine bismuth, cadmium, and indium by the standard addition technique, liquid standards being added to weighed portions of the solid samples.

Determine nickel, chromium, lead, and silver by atomizing samples solutions in the Varian-Techtron tube furnace, the measurements being made by the standard addition technique.

Determine cadmium, chromium, cobalt, copper, and manganese and zinc in the air–acetylene flame using the standard addition method.

The operating conditions for the spectrophotometers and graphite furnaces are listed in Table XVIII.

Determination of Cadmium, Copper, Nickel, Zinc, Cobalt, and Silver in Rocks and Sediments (87)

Comment on the Method

Nonspecific absorption can be a serious problem when very low levels of a metal are to be analyzed in a complex matrix. This method allows the dithizone extraction of zinc, copper, nickel, cobalt, cadmium, and silver from the sample matrix prior to their determination. Cadmium and zinc are back extracted into dilute hydrochloric acid solutions, the copper, nickel, cobalt, and silver are determined after destruction of the organic matter.

The main procedure is not satisfactory for calcium-rich oceanic sediment. A separate method is given for this latter sample type. The methods were tested on USGS reference rocks with satisfactory results.

Reagents and Equipment

Analar reagents were used unless otherwise stated.

Buffer solution.   Add 1.3 g of ammonium chloride and 4.4 ml of ammonia (1:49) to water and dilute to 500 ml.

Chloroform. Purify by first shaking with diluted buffer solution (1:5) and then twice with hydrochloric acid, using about 500 ml of aqueous solution to each 100 ml of chloroform.

Dithizone solutions. First, recrystallize the dithizone by passing a current of filtered air into a nearly saturated solution of dithizone in chloroform at 40°C, and wash the precipitate with a small volume of carbon tetrachloride. Make a 1000 μg/ml solution in chloroform daily.

Procedure: Most Samples

Weigh 1 g of sediment into a PTFE crucible (~100 ml capacity), add 10 ml of hydrofluoric acid (Aristar grade), 5 ml of sulfuric acid (lead-free grade), and 0.5 ml of perchloric acid. Cover and leave on the sandbath at 100–150°C overnight. Remove the cover and leave open for 1 hr. Transfer the contents to a 150 ml tall-form beaker. Heat the beaker on a hot plate until white fumes are given off. Cool, add 0.5 ml of perchloric acid, and heat the mixture until dense fumes are evolved and most of the color arising from organic matter has disappeared; do not evaporate to or near dryness. Cool, add water to make the volume 60–80 ml, and heat on the hot plate until a clear solution is obtained. Cool, add 5 ml 50% citric acid and about 10 ml ammonia (Aristar grade). Cool again, transfer to a 100–150 ml separating funnel and adjust the pH to 8 with ammonia or 2 N hydrochloric acid (indicator papers). Add 5 ml of 1000 μg/ml dithizone solution, shake for 1 min, and run the organic layer into a second 100–150 ml separating funnel, adjust the pH of the solution to 9.5, and repeat the extraction. Extract once more, this time with 5 ml of 200 μg/ml dithizone solution. If the dithizone still changes color, repeat the extractions until it does not. Wash the combined dithizone extracts with 50 ml of buffer solution. Run the dithizone layer into a third 100–150 ml separating funnel, wash the buffer solution with 5 ml of chloroform, and add the washings to the dithizone extracts. Add 50 ml of 0.2 M hydrochloric acid to the dithizone extracts, and shake vigorously for 1 to 2 min. Run the organic layer into a 150 ml tall-form beaker, wash the aqueous portion with 5 ml of chloroform, add the washings to the dithizone solution, and run the aqueous layer into the original 150 ml tall-form beaker.

Evaporate the aqueous solution to 10–15 ml, add 0.5 ml of perchloric acid and 1.5 ml of nitric acid, and continue the evaporation to dryness. Rinse the wall of the beaker with 10–15 ml of distilled water, and evaporate to dryness again. Dissolve the residue in a small volume of 2 M hydrochloric acid, transfer to a 5 ml volumetric flask (or a larger one if high concentrations of cadmium are suspected), fill to the mark with 2 M

hydrochloric acid, and determine cadmium against standard cadmium solution in 2 M HCl at 2288 Å using a suitable scale expansion. Determine zinc against similarly prepared standard solutions at 2138 Å, turning the burner around 90° if necessary.

Evaporate the organic extract to dryness. Add 3 ml of perchloric acid and 9 ml of nitric acid. Digest and evaporate to dryness on a hot plate. Repeat with 2 ml of perchloric acid. Wash down the beaker with 10–15 ml of distilled water, and evaporate to dryness. Take up the residue and transfer to a 5 ml volumetric flask, using 2 M hydrochloric acid. Determine nickel at 2320 Å, cobalt at 2407 Å, silver at 3281 Å, low concentrations of copper at 3247 Å, and high concentrations of copper at 3274 Å.

Procedure: Calcium-Rich Oceanic Sediment

Weigh 1 g of sediment into a PTFE beaker (~100 ml capacity), add 20 ml constant boiling hydrochloric acid, and digest on the sandbath for 2 to 4 hr. Transfer the contents to a centrifuge tube and spin. Add the supernate to a 150 ml tall-form beaker and evaporate slowly to dryness. Transfer the residue from the centrifuge tube, add 5 ml hydrofluoric, 2 ml sulfuric, and 0.5 perchloric acids, cover, and leave on the sandbath at 100–150°C overnight. Remove the cover and leave open for 1 hr. Add the contents to the 150 ml beaker containing the residue of the hydrochloric acid extract. Heat the beaker on a hot plate until white fumes are given off and then cool. Add 0.5 ml perchloric acid and heat the mixture until dense fumes are evolved and most of the color arising from organic matter has disappeared. Add 50–60 ml of water and 20 ml constantly boiling hydrochloric acid and heat. Add further 5 ml portions of hydrochloric acid until a clear solution is obtained (total addition 30–40 ml), cool, add 10 ml 50% citric acid solution, and 20–25 ml ammonia (Aristar), and proceed as previously described.

Determination of Trace Metals in Rocks and Minerals (88)

Comment on the Method

The following procedure is very simple, rapid, and straightforward. However, it can only be used for concentrations greater than about 5 μg/g in the sample.

For cadmium, manganese, zinc, iron, cobalt, copper, nickel, and chromium, a hydrofluoric–sulfuric acid dissolution may be used. When lead is to be analyzed, sulfuric acid must be replaced by perchloric acid. If silver is of interest, the hydrochloric acid should be avoided.

The following procedure has been tested using USGS standard rocks. Even considering uncertainties in the accepted values for these standards, it has been found that the method gives satisfactory results.

Reagents and Equipment

A Perkin-Elmer 305B unit was used with an air–acetylene flame. The resonance wavelengths, slit widths, and flame conditions were similar to those recommended by the manufacturer. In the case of chromium, a lean flame, which minimized interferences, was used.

Standard solutions are prepared from pure metals or analytical grade metal salts. Dilute working solutions are made by dilution of the above to the appropriate concentration range so that the final acid concentration is 1%. Work in the author's laboratory suggests that multielement dilute working solutions may be used.

Procedure

Weigh an appropriate-sized sample, usually 0.5–1.0 g, into a 150 ml Teflon dish. Moisten with a few drops of water. Based on a 1.0 g sample, add 25 ml hydrofluoric and 1 ml sulfuric acid (substitute perchloric and nitric acids if lead is to be analyzed). Heat to dryness using medium heat on a hot plate. Add 1 ml hydrochloric acid (use nitric acid if silver is to be analyzed) and heat on medium heat for 5 min. Dilute with 10 ml of water. Filter the sample into a 25 ml flask. Wash the filter paper well with dilute acid. Dilute to volume, cap, and shake well. Dilutions of this stock may be necessary. The diluted solutions should have 1% acid content.

Samples are run on the atomic absorption unit using an air–acetylene flame and the instrumental parameters recommended by the manufacturer. A very lean flame is essential for nickel and chromium. It is important to use the background corrector, even for elements such as chromium and copper, whose resonance wavelengths are above 3000 Å.

Determination of Metals in Lake Sediments (89)

Comment on the Method

A very wide range of elements (beryllium, calcium, cadmium, cobalt, chromium, copper, iron, potassium, lithium, magnesium, manganese, molybdenum, sodium, nickel, lead, strontium, vanadium, and zinc) can be done by this procedure. Hydrofluoric along with perchloric and nitric acids are employed for decomposition. Digest is carried out in a closed container such as a Parr digestion bomb. This prevents loss of chromium as volatile chromyl chloride (CrO_2Cl_2). Experience in the author's laboratory indicates that similar results, except for chromium, can be obtained using Teflon dishes if the acid volumes are increased by a factor of 2 and the sample fumed to dryness. Recoveries on the lake sediments tested were between 91 and 101%, depending on the element.

Reagents and Equipment

A Parr 4745 acid digestion bomb (Parr Instrument Company, Moline, Illinois) was used for the decompositions.

The atomic absorption spectrometer (Perkin-Elmer Model 403) was fitted with either a Boling (three-slot) air–acetylene burner or a nitrous oxide burner, and used with an A-25 Varian strip-chart recorder. All instrumental settings, including wavelength, slit, and grating, were those recommended by the manufacturer. Cadmium, cobalt, chromium, copper, iron, potassium, lithium, manganese, nickel, sodium, lead, and zinc were determined by using an air–acetylene flame. A Boling three-slot burner was used for all metals except sodium and potassium for which a 2 in. air–acetylene burner was used. Aluminum, barium, beryllium, calcium, magnesium, molybdenum, strontium, and vanadium were determined with a nitrous oxide–acetylene flame. The burner position for all the metals was parallel to the hollow-cathode light path except for calcium, potassium, magnesium, and sodium when it was held at 90° to reduce the sensitivity.* Perkin-Elmer hollow-cathode lamps were used at lamp currents specified on the lamps.

High-purity certified reagents were used for all analyses. All standards were prepared from certified atomic absorption reference standards (Fisher Scientific Co.).

Procedure

Thaw frozen sediment samples and air-dry at room temperature. Crush the dried sample to a fine powder and take a representative sample of about 2 g. Further grind this sample until a very fine powder passing through a 270 mesh sieve is obtained. Transfer the prepared sample (100 mg) to the Teflon cup, and add 4.0 ml of concentrated nitric acid, 1.0 ml of perchloric acid (60%), and 6.0 ml of hydrofluoric acid. Seal the bomb and heat for 3.5 hr at 140°C. After cooling, transfer the contents of the Teflon cup, quantitatively, to a 125 ml of polypropylene bottle containing a solution of 4.8 g of boric acid in about 30 ml of

---

* It may be best to dilute some of the sample solution further to avoid noisy signals often obtained in this way.

deionized water to dissolve the precipitated metal fluorides. Transfer the solution to a 100 ml volumetric flask, make up to volume, and store in a 125 ml polypropylene bottle.

Standards are made to contain 4.0% (v/v) nitric acid, 1.0% perchloric acid (60%), 6.0% hydrofluoric acid, and 4.8% of boric acid. A blank sample is prepared similarly to the unknown samples and is taken through all the steps involved.

Determine beryllium, cadmium, chromium, cobalt, copper, lithium, magnesium, lead, and zinc by direct aspiration of the main sample solution. Determine copper by making all solutions and standards 2% (w/v) in ammonium chloride to overcome interference from iron. For molybdenum and chromium, prepare all solutions and standards with 500 mg Al/liter (as $AlCl_3$). Determine manganese, iron, and aluminum by diluting the main sample solution by a factor of 10 and analyzing by direct aspiration. For the determination of barium, calcium, and strontium, all sample and standard solutions must contain 3000 mg Na/liter (as NaCl), in order to remove the ionization interference in the nitrous oxide–acetylene flame. To prevent ionization interference in analyses for sodium and potassium, all solutions should contain 100 mg Cs/liter (as CsCl).

The acid matrix used in the decomposition could suppress the signal of some metals. Nitric acid is known to suppress the signal for iron. Therefore, all reagents added to the samples are added to the standards in the same amounts.

Determination of Trace Elements in Brown Coal (90)

Comment on the Method

The following 20 elements are covered by the procedure: lithium, beryllium, vanadium, chromium, manganese, cobalt, nickel, copper, zinc, germanium, arsenic, strontium, molybdenum, silver, cadmium, antimony, barium, mercury, lead, and bismuth. A schematic of the procedure is given in Fig. 29. Mercury is analyzed by a cold-vapor technique. Lithium, beryllium, vanadium, chromium, manganese, cobalt, nickel, copper, zinc, molybdenum, cadmium, and lead are done by flame atomic absorption using a standard addition approach for calibration. Solvent extraction using sodium diethyldithiocarbamate with di-isobutyl ketone solvent is used with standard addition flame atomic absorption for vanadium, cobalt, nickel, molybdenum, cadmium, and lead. Silver, strontium, and barium are done by conven-

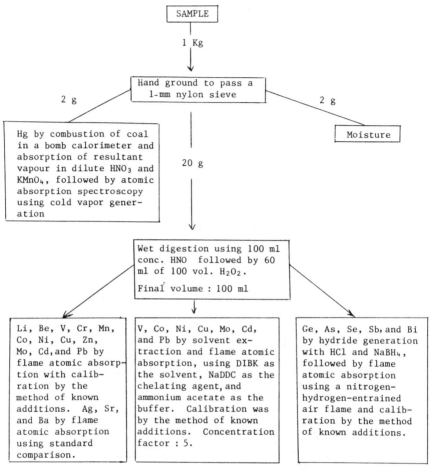

Fig. 29. Schematic diagram for the determination of trace elements in brown coal (90).

tional flame atomic absorption. The remaining elements are analyzed by air–hydrogen flame atomic absorption using hydride generation. Calibration for these latter elements is done by the standard addition method.

Accuracy of the method was estimated by the analysis of NBS Orchard Leaves SRM 1571. Satisfactory agreement was demonstrated for most elements. Lithium and bismuth, not certified by NBS, could not be tested for accuracy. Selenium yielded poor precision and was not included in the method. The original authors stress the need for nonspecific background correction in all determinations.

Reagents and Equipment

The atomic absorption instrumentation consisted of a Varian-Techtron AA6 equipped with a model BC-6 simultaneous background corrector, an M80 automatic gas control unit, and 5 and 10 cm burners; a Perkin-Elmer Model 460 double-beam unit, equipped with automatic gas control, simultaneous background correction with a deuterium arc lamp, 5 and 10 cm burners, and Varian-Techtron hollow-cathode lamps. An LDC uv monitor 1205 equipped with a long-path gas cell was used for mercury determinations, and a Varian-Techtron Model 64 vapor generation kit was used for the generation of volatile hydrides. Other instrumentation consisted of a Gallenkamp autobomb adiabatic bomb calorimeter, a Brabender moisture tester, and a Phillips PW 9418 pH meter. Containers were cleaned by soaking them in 1 M nitric acid for at least 24 hr. Reagents used were A.R. or better quality. Stock standards were either prepared by dissolving the appropriate metal in concentrated nitric or hydrochloric acid, or were purchased from BDH or Merck. An arsenic(III) standard was prepared from arsenious oxide.

Procedure

Block samples were taken either from the cleaned coal face or from core drillings, and immediately sealed in plastic bags. At the laboratory, the outer layers from each block were removed using a large Teflon scraper. The samples were then crushed in unglazed wedgewood mortars to pass a 1 mm nylon sieve, with care being taken to ensure thorough mixing, and then stored in white polyethylene containers.

Particular care is necessary when preparing solid samples for trace metal analysis. Contamination can occur with the use of mechanical grinders. In the present study, materials used during the preparation of the coal samples for analysis were chosen on the basis that they cause minimal contamination of samples. Hand grinding, though tedious, was thought to be the safest approach.

Weigh samples of coal (20 g) into 400 ml (plus) tall glass beakers fitted with watchglasses. At the same time, seal coal samples (2 g) in plastic vials for moisture determination. Add concentrated nitric acid (100 ml) to each beaker and allow the digestion to proceed unaided for 30 min. Gently heat the samples to boiling point and reduce the volume to about 25–30 ml. After cooling add 100 volume hydrogen peroxide (20 ml) to each sample. Allow the samples to stand until the reaction subsides. Reduce the volume to 25–30 ml by heating and repeat this process twice more using further 20 ml aliquots of hydrogen peroxide. Make samples up to 100 ml. Blank samples were prepared in the same way.

Some elements are potentially volatile during wet digestion; however, the results show that no losses occurred during the wet digestion of the three coal samples tested. One notable result is the loss of antimony from the blank during digestion, which could potentially cause errors. However, no such loss occurred from the coal digests.

To 20 ml volumes of the wet digestion samples, add 0, 1, and 2 ml aliquots of standard solution. The standard should contain vanadium, molybdenum (50 mg/liter), chromium, manganese (40 mg/liter), and lithium, germanium, cobalt, nickel, copper, zinc, cadmium, and lead (20 mg/liter). Prepare blanks in the same way. Determine silver by standard comparison with 1 and 2 mg/liter standards. Determine strontium and barium after samples, blanks, and suitable standards are made 5000 mg/liter in potassium chloride. Use instrumental conditions as recommended by the manufacturers, except for lead, which is determined using the 2833 Å resonance line.

Silver chloride precipitation after digestion can interfere with the determination of silver. However, additions of silver to coal digests compare well with additions to blanks and deionized water, indicating that no precipitation is occurring.

For solvent extraction, three 30 ml volumes of each sample are transferred to 500 ml separation funnels, and 1 ml of standard is added to one aliquot. The standard contains vanadium, molybdenum, lead (5 mg/liter), and cobalt, nickel, copper, and cadmium (2 mg/liter). Buffer sample solutions to a pH of 4.5 using 50% w/v ammonium acetate, which was previously purified. Add 10 ml of purified 5% w/v sodium diethyldithiocarbamate and 6 ml of di-isobutylketone and immediately shake the extraction flasks for 1 min, after which the organic phase is collected for analysis. Prepare blanks in the same way.

Sodium diethyldithiocarbamate instead of ammonium pyrrolidine dithiocarbamate is used because the high iron levels in the digest cause substantial interference with the latter. However, because solvent extraction using sodium diethyldithiocarbamate di-isobutylketone often contributes a substantial nonatomic signal, simultaneous background correction is necessary, with the exception of vanadium and molybdenum.

Use instrumental conditions as recommended by the manufacturers except for lead (2833 Å). Reduce the fuel flow rate to compensate for the fuel effect of the ketone. Deionized water or 5% v/v HNO_3 is periodically aspirated to prevent the nebulizer from clogging.

Determine vanadium as soon as possible after extraction because the complex is unstable in di-isobutyl ketone. Vanadium can generally exist in solution in the 4 and 5 oxidation states. Only the 5 state is found to respond to solvent extraction. However, it is likely that the oxidative digestion of the coal will convert all vanadium to its highest oxidation state. Undissolved silaceous matter in the coal digests does not interfere substantially with the extraction.

For mercury, 2 g samples of coal are weighed into the combustion cap of a bomb filled with 50 ml of 0.5% potassium permanganate. The bomb is charged with oxygen to a pressure of 20 atm and fired in the standard way. After 10 min open the bomb and transfer the contents to a 100 ml standard flask and make up to volume. Just prior to analysis, add sufficient 20% hydroxylamine hydrochloride solution to the standard flask to discolor the potassium permanganate and dissolve the precipitated manganese dioxide. This step is important because some mercury may be absorbed onto the manganese dioxide particles, causing low results. Prepare sample blanks in the same way using an empty combustion cup on the bomb. Initial conditioning of the bomb is achieved by repeated blank firings. Standards are prepared by adding 0, 5, 10, 20, and 40 ml of a 40 μg/liter mercury solution (preserved with nitric acid and potassium dichromate) to 50 ml of potassium permanganate solution and diluting to 100 ml. Analyze sample volumes of 10 ml by cold vapor generation using 1 ml of 20% stannous chloride in 10% v/v HCl.

For the determination of germanium, arsenic, antimony, and bismuth make a 5 ml volume of coal digest the appropriate strength in hydrochloric acid, as given in Table XIX, and place in the test tube part of the vapor generation kit. For arsenic and antimony determination, 2 ml of 25% potassium iodide is also added to each solution. Fit the test

Table XIX

Experimental Conditions for Chalconides

Element	HCl concentration (M)	Other conditions	Amount that gives 0.100 absorbance peak in coal solution matrix (μg)
Ge	1.5		5
As	3	KI added	0.2
Sb	3	KI added	0.2
Bi	3		0.25
Se	6	Sample evaporated to dryness in the presence of NaOH and CaCl$_2$, redissolved in NH$_4$Cl and HCl, and heated on a steam bath	0.4

tube head firmly and inject 5 ml of 5% w/v sodium borohydride rapidly through the septum from a syringe. In order to obtain reproducible results, it is important to inject the borohydride solution rapidly and to ensure that it impinges directly into the solution in the test tube without making contact with the walls of the tube. The generated gases are swept through the auxiliary oxidant inlet into a nitrogen–hydrogen–entrained air flame, and the resultant peak is recorded.

For each element blanks are prepared in the same way as samples, and suitable amounts of germanium(IV), arsenic(III), antimony(IV), or bismuth(III) are added to samples and blanks for calibration purposes. Instrumental conditions are as recommended by the manufacturers and simultaneous background correction with a deuterium arc is used for all elements.

Determination of Mercury in Coals (91)

Comment on the Method

Oxidizers, e.g., chromium trioxide and potassium permanganate, other than hydrogen peroxide were found to be unsatisfactory for the decomposition of coal. A 30% hydrogen peroxide–concentrated sulfuric acid solution was chosen.

The method was tested on NBS Coal #1630 certified at 0.15 ± 0.01 ppm (accepted range 0.14–0.16 ppm). Good reproducibility was obtained except on some samples containing high mercury levels.

Reagents and Equipment

Hydrogen peroxide. Baker analyzed reagent, 30% solution, stabilized.

Hydroxylamine hydrochloride. Dissolve 10 g of Baker analyzed reagent in 100 ml deionized water; prepare fresh daily.

Stannous chloride solution. Dissolve 10 g of Mallinckrodt analytical reagent stannous chloride in 10 ml Baker analyzed hydrochloric acid and dilute to 100 ml with deionized water. A few grains of tin shot stabilize the solution.

Mercury standard solution. Pipet 100 μl of Utopia Mercury standard (1000 μl/ml) and 100 μl of concentrated nitric acid into a 100 ml volumetric flask and dilute to volume with deionized water to produce a mercury standard.

Silverwool. J. T. Baker Company silver wool for microanalysis was used.

The Utopia mercury kit glassware assemblage (Utopia Instrument Co., P.O. Box 683, Joliet, Illinois) is used for generating the mercury metal, and the mercury is retained by amalgamation on a 1.5 silver wool plug inserted into a 15 cm Vycor tube (1 cm diam). The glassware assemblage consists of a 250 ml bubbling cylinder, which is controlled by a bypass Teflon stopcock and which has a ground-glass ball joint sample introduction port; the generating cylinder is followed by another 250 ml cylinder, which serves as a trap for water droplets. Argon flows through the cylinder at a rate of approximately 4 liters/min.

Procedure

Weigh 0.20 g of 100 mesh coal into a 125 ml Pyrex Erlenmeyer flask and add 10 ml concentrated sulfuric acid. Cover the flask with a watchglass and heat at low heat on a hot plate (~100°C). Add 30% hydrogen peroxide dropwise until the mixture becomes clear or until about 4 ml of peroxide has been added. Add 3 ml of 10% hydroxylamine hydrochloride solution, dropwise, to destroy excess hydrogen peroxide. The reaction is somewhat violent generating heat and nitrogen oxide fumes. Carry a blank (excluding only the coal) through the whole procedure.

To generate mercury from a coal solution, the following procedure is utilized.

Place about 50 ml deionized water into the gas-bubbling cylinder of the mercury glassware assemblage and add 5 ml of 10% stannous chloride solution. Install the cylinder onto the apparatus and bubble argon through the solution for one minute with sample port open (this purges mercury from the reductant). Turn the stopcock to arrest the flow of argon.

Add the sample solution through the sample port and wash it into the cylinder with deionized water from a squirt bottle. Close the port. Turn the argon flow to bypass the sample cylinder.

When the Vycor tube is cool to the touch, turn the stopcock to allow argon to flow through the sample solution. Let bubble for 3 min. During this time mercury is being generated, transported, and deposited onto the silver wool. Turn the stopcock to allow argon to bypass sample solution.

Place a Meker burner 14 cm under the Vycor tube and heat the tube, releasing mercury from the silver. Sweep the mercury vapor away by the argon stream and pass through a glass wool trap to remove water

droplets and into a 10 cm quartz-windowed cell placed in the beam of a Perkin-Elmer Model 403 spectrophotometer. Use an Intensitron mercury hollow-cathode lamp selecting the 2537 Å line. Record instrument response on a 10 mV Houston Instruments Omniscribe strip-chart recorder. Full scale of the recorder corresponds to about 0.08 absorbance units (recorder full scale is set on the 403 to 0.25 abs and scale expansion at 3 from concentration dial). Full scale corresponds to about 50 ng of mercury. Cool the Vycor tube by the argon flow (after the burner is turned off) and by application of a wet towel to the tube (after some initial cooling).

Determination of Lead in Coal and Coal Ashes (92)

Comment on the Method

Suprisingly, an ashing temperature of 900°C is used. At this temperature there is a danger that lead will be lost. However, results on standard reference samples do not indicate a problem.

A nitric acid–hydrofluoric acid mixture in a pressure bomb was found to be the best decomposition medium. Analysis is done in a graphite furnace. Both acids suppress the lead absorbance so a minimum quantity of the mixture should be employed. The accuracy and precision of the method was tested using NBS Coal and Fly Ash. Results show satisfactory agreement with accepted values. Precision was generally better than 10%.

Reagent and Equipment

A Perkin-Elmer Model 503 double-beam atomic absorption spectrophotometer was used with an Intensitron hollow-cathode lamp, a deuterium background corrector, an HGA 2100 graphite furnace and a Hitachi Perkin-Elmer Model 056 recorder. Coal and coal ash samples were decomposed in an acid digestion bomb, consisting of a Teflon crucible encased in a metal body with a screw cap.

A Branson ultrasonic cleaner consisting of an ultrasonic generator Model LG 150 and an ultrasonic tank-type transducer LTH 60 was used to remove fly ash from the filter paper.

Procedure: Coal and Coal Ash

Crush and homogenize the coal samples, and ash them in platinum crucibles in an electric furnace at 900°. Transfer about 200 mg of coal or about 30 mg of coal ash to an acid digestion bomb, add a mixture of 3 ml of 14 M nitric acid (Merck suprapur) and 4 ml of 30 M hydrofluoric acid, and heat at 230°C for 6 hr. After cooling, dilute the solution to 100 ml with distilled water and store in polyethylene containers. The final solution contained approximately 0.42 M nitric acid and 1.2 M hydrofluoric acid.

Prepare a lead stock solution from reagent-grade lead nitrate and dilute to concentrations varying from 10 to 400 μg/liter, with a solution whose acidity is similar to that of the unknown samples.

Procedure: Fly Ash

Collect the fly ash samples on Whatman No. 41 filter papers. Remove the collected dust from the filter by ultrasonic vibration in a 0.1 M nitric acid solution for aerosol samples. Use dust fractions corresponding to about 10 mg of fly ash. Prepare standard solutions of lead nitrate dissolved in 0.1 M nitric acid, in concentrations ranging from 0.25 to 5 mg/liter.

Absorption Measurements

Use the following optimal apparatus settings: lamp current, 10 mA; slit, 10 Å; drying, charring, and atomization temperatures 100, 400, and 2700°C, respectively. Inject 10–75 μl of sample solution into the graphite tube with Eppendorf micropipets, and measure at the lead 2833 Å line. Use the deuterium background corrector to compensate for scattering and broadband absorption.

After subtraction of the blank value for the reagents, calculate the concentrations of lead in the coal, coal ash, and fly ash from the appropriate calibration curves.

Determination of Iron in Coals (93)

Comments on the Method

Iron is analyzed in the sample solution obtained by using ASTM method D 2492. By the ASTM method the iron was determined by titration. Results obtained on a highly volatile carbon bituminous coal by the ASTM titration and the proposed atomic absorption method are comparable.

Reagents and Equipment

A Perkin-Elmer 303 unit was employed. Reagent grade hydrochloric acid was used to prepare a 2:3 v/v dilution. A 1:7 nitric acid solution was made from reagent grade acid.

Procedure

Grind the coal to 60 mesh. The coal is then subjected to hydrogenation treatment. Excess solvent is removed by distillation.

Extract samples with 100 ml of boiling 2:3 v/v hydrochloric acid for 0.5 hr. Filter into 1 liter flasks. Treat the residue with boiling 1:7 nitric acid v/v for 0.5 hr. Filter into the 1 liter flasks. Dilute to volume. Standards are prepared to contain the same concentration of acids as the samples.

Run samples and standards on the atomic absorption using an air–acetylene flame and the operating conditions recommended by the manufacturer.

Analysis of Iron Ores (94)

Comment on the Method

The following procedure is suitable for the analysis of aluminum, barium, calcium, copper, manganese, sodium, silicon, titanium, and zinc in limonite, hematite, and siderite. Decomposition with a mixture including hydrofluoric acid is done in a pressure vessel at 145–155°C. Boric acid solution is added subsequently to complex excess fluoride. A procedure is also given for iron, but in keeping with the premise that atomic absorption is not suitable for most major element analyses, this has not been included here.

Results have been obtained on standard iron ores for calcium, aluminum, magnesium, manganese, titanium, and silicon. These agree well with accepted values.

Reagents and Equipment

Digestion vessels (Bodenseewerk Perkin-Elmer autoclave-1 and autoclave-2, which are similar to the autoclave developed by Langmyhr and Paus (95), were used for the digestions.

The body of the autoclave is made of aluminum, and the inner vessel with cover is made of PTFE, which is resistant to temperatures up to 250°C. The sample plus acid, sealed in such a vessel, is heated to a maximum of 160°C. An inner pressure of 6 atm (autoclave-1) or 45–50 atm (autoclave-2) is permitted. Overheating causes a safety valve inside the aluminum cover to open and causes a loss of elements. Figure 30 shows the autoclave-2. It was found that adding another plate spring to the row of safety valve springs increases the maximum pressure to about 8 atm and further improves the performance of the autoclave-1.

All measurements with the flame technique were made with a Perkin-Elmer Model 403 double-beam atomic absorption spectrometer

Fig. 30.   Autoclave-2, Bodenseewerk Perkin-Elmer Ltd.

equipped with a deuterium background corrector and a Model 056 recorder.

A single decomposition of a sample was made and a single determination in 100 AVERAGE mode was used, the average of at least three such readings.

For the flameless technique, a Perkin-Elmer Model 300S atomic absorption spectrometer in combination with an HGA 74 graphite furnace, a corresponding background corrector, and a recorder were used.

Eppendorf micropipets were used for injecting the samples into the graphite tube of the HGA 74.

All chemicals were of analytical grade (Merck).

Determinations of aluminum were carried out in a nitrous oxide–acetylene flame. Amos and Thomas (96) found that the presence of calcium, zinc, copper, lead, magnesium, sodium, phosphate, or sulfate did not influence the sensitivity of measurement. Only iron and hydrochloric acid in concentrations higher than 0.2% decrease sensitivity. In order to control ionization interferences, samples and standards were diluted with a 200 μg K/ml solution and were measured with a nitrous oxide–acetylene flame. The optimum working range was 10–150 μg Al/ml; the sensitivity was about 1.3 μg Al/ml for 1% absorption.

Both air–acetylene and nitrous oxide–acetylene flames are suggested for the determination of calcium. To avoid interference from silicon, aluminum, phosphate, and sulfate, either lanthanum or strontium, usually a 0.25% strontium solution, as chloride must be added. To overcome ionization interferences, Langmyhr (95) suggests an addition of potassium.

The interferences in an air–acetylene flame were obvious. A rather high content of silicon in the samples could be the reason for this. However, with the nitrous oxide–acetylene flame good agreement with the certified values was obtained. The sensitivity was about 0.09 μg Ca/ml for 1% absorption.

The determination of copper is usually possible without any difficulty. At the 3247 Å nm line the sensitivity is about 0.15 μg Cu/ml for 1% absorption. The optimal working range was 2–20 μg Cu/ml.

For potassium, in order to reduce possible contamination from the atmosphere during determination, a three-slot burner head was used. This is advisable only when determinations of very low potassium concentrations are performed. For optimum sensitivity Gatehouse and Willis (97) recommend a low-temperature acetylene-rich flame. The optimum working range was found to be 1–10 μg K/ml.

If aluminum is present, a nitrous oxide–acetylene flame, rather than the use of a realeasing agent, is recommended in the determination of magnesium. Aqueous standards were prepared with the same amount of aluminum as in the samples. On the other hand, addition of 1500 μg Sr/ml controlled interferences from up to 100 μg Al/ml. Langmyhr and Paus (95) suggested an addition of 10 ml of 10 μg K/ml solution to each sample. The optimum working range for the air–acetylene flame is 0.1–2 μg Mg/ml, and the sensitivity is 0.01 μg Mg/ml for 1% absorption.

In the determination of manganese, interferences from silicon in an air–acetylene flame could be expected. Interferences exceeding the analytically acceptable limit were not noted in this application. The working range is 2–20 μg Mn/ml, and the sensitivity is 0.10 μg Mn/ml for 1% absorption.

The optimum working range for sodium in an air–acetylene flame is 0.3–3 μg Na/ml. The sensitivity is about 0.15 μg Na/ml. A three-slot burner head was used.

A nitrous oxide–acetylene flame is required for silicon analysis and the optimum working range is 20–200 μg Si/ml. The sensitivity is about 1.5 μg Si/ml for 1% absorption.

Zinc is dissociated very well in air–acetylene flame, and the sensitivity is 0.25 μg Zn/ml for 1% absorption. The optimum working range is 0.2–3 μg Zn/ml.

The elements with concentrations below the detection limits of the flame technique can be determined with the graphite furnace. The aliquot injected into the graphite furnace was always 25 μl. Instrumental settings are shown in Table XX.

During atomization, an internal flow of argon (50 ml/min MINI FLOW) was always used. Nonspecific absorption was corrected using a built-in deuterium background corrector.

Procedure

Transfer a representative sample of 0.20 g, ground to less than 0.070 mm into the PTFE vessel, and add 0.750 ml concentrated hydrochloric acid, 0.250 ml concentrated nitric acid, and 5.0 ml hydrofluoric acid

Table XX

Atomic Absorption Instrumental Conditions

Element	Wavelength (Å)	Slit (Å)	Flame
Al	3093	7	N_2O/C_2H_2
Ba	5536	2	N_2O/C_2H_2
Ca	4227	7	N_2O/C_2H_2
Cu	3247	7	air/C_2H_2
Fe	2483	2	air/C_2H_2
K[a]	7665	20	air/C_2H_2
Mg	2852	7	air/C_2H_2
Mn	2795	2	air/C_2H_2
Na[a]	5890	7	air/C_2H_2
Si	2516	2	N_2O/C_2H_2
Ti	3643	2	N_2O/C_2H_2
Zn	2139	7	air/C_2H_2

[a] Three-slot air/C_2H_2 burner head.

(38%). After controlled heating at a temperature of 145–155°C and simultaneous magnetic stirring for about 30 min, cool the sample vessel with tap water. Add 50 ml of saturated solution of boric acid and if a clear solution has not been obtained, heat the open vessel with solution while stirred until the solution is clear. After transferring to a plastic 100 ml volumetric flask and diluting with water, the solution is ready for analysis by atomic absorption spectrometry.

If it is not possible to obtain a clear solution, the operation has to be repeated with a sample ground to a finer powder or with prolonged heating.

Determine the elements with the flame or furnace using the parameters listed in Tables XX and XXI, respectively.

Determination of Arsenic in Geological, Biological, and Environmental Samples (98)

Comment on the Method

This method requires the least amount of hydride generation apparatus (one disposable syringe) of all procedures published. It is applicable to a very wide range of sample types. The main interference problem is with nickel and thus it may seem strange to use nickel crucibles for fusions. However, these crucibles are the least costly of the suitable types. If the precautions outlined in the procedure are taken, no problem with crucible-generated nickel will be encountered.

The fusion and acid decomposition procedures given were subjected to exhaustive testing. Most of the procedural steps to be listed were

Table XXI

Instrumental Settings for HGA-74

			HGA settings (°C-sec)			
Element	Wavelength (Å)	Slit (Å)	Drying	Thermal pretreatment	Atomization	Optimum range (ng)
Co	2407	2	100–30	1150–30	2660–15	2–20
Cr	3579	7	100–30	1150–30	2660–15	0.5– 5
Ni	2320	2	100–30	1200–30	2660–15	2–20
Pb	2833	7	100–30	550–30	2050–10	1–10
Sb	2176	2	100–30	1000–30	2660–15	2–20[a]
V	3183	7	100–30	1700–30	2660–20	10–80

[a] Standards were prepared in 0.03 N H_2SO_4.

established on the basis of extensive investigation in the authors' laboratories and should be adhered to very closely.

The fusion procedure results in decomposition of most siliceous materials. It should be used when total arsenic is required even with samples containing only small amounts of silicate. The fusion should also be used on samples likely to contain large amounts of materials that form sulfate precipitates (e.g., barium, strontium, calcium, and lead).

Steel wool must be used to clean crucibles. Other methods such as acid washes cause too much nickel to be released in the next fusion. If acid cleaning is periodically necessary, a blank fusion should be done before reusing the crucible. Fusion time is not critical. Fifteen minutes is optimum for decomposition at 550°C without causing excess release of nickel.

The acid digestion procedure is more rapid and can be employed for organic samples devoid of siliceous material. This method is also useful when only mineral-acid-extractable arsenic is required.

There is a wide variation in arsenic content in hydrochloric acid from various commercial hydrochloric acid suppliers. It is important to select acid with a low arsenic content. By extensive testing Fisher ACS hydrochloric acid was found to be satisfactory.

The reader may suspect that the hydride generation procedure is subject to large operator error. The author finds that the technique can be easily mastered with 1 to 2 hr practice. Subsequently a 2–5% coefficient of variation is commonly obtained.

The borohydride–sample mixture in the syringe must be shaken for 20 sec. This allows the hydride reaction to proceed to completion. Maximized arsenic signals, minimum interference, and better reproducibility are thus obtained.

The procedure gives good precision, ±5%. It has been tested on biological, environmental, and geological standard samples with good results. Table XXII lists potential interferences and the concentrations at which they cause problems.

Reagents and Equipment

Arsine gas was generated in a 50 cm^3, R-1314 (Becton and Dickinson) syringe fitted with a Yale No. 20 stainless steel needle. Each syringe lasts for 100–200 injections.

A 3.18-cm length of 1.91-in.-diam thick-walled rubber tubing is placed between the ribs of the syringe to act as a stopping mechanism during injection of the hydride. This allows quick injection while preventing introduction of the liquid into the atomizer.

The procedure can be used with most atomic absorption equipment.

Table XXII

Interferences in As Determination

Cation	Solution concentration of cation (μg/ml)	Concentration equivalent for 1 g sample in 100 ml solution
Ca^{2+}	4000	40%
Al^{3+}	4000	40%
Mg^{2+}	4000	40%
Na^+	1000	10%
K^+	1000	10%
Li^+	1000	10%
Fe^+	750[a]	7.5%
Pb^{2+}	200[a]	2%
Zn^{2+}	200	2%
Mn^{2+}	200	2%
Ba^{2+}	200	2%
Cr^{3+}	100[a]	1%
Co^{2+}	100[a]	1%
Cd^{2+}	100[a]	1%
Cu^{2+}	10[a]	1000 ppm
Ni^{2+}	6[a]	600 ppm
Ag^+	5	500 ppm
Ge^{4+}	1[a]	100 ppm
Sb^{3+}	0.06[a]	6 ppm
Sn^{2+}	0.08[a]	80 ppm
Bi^{3+}	0.6 [a]	60 ppm
Se^{2+}	0.04[a]	4 ppm
Te^{2+}	0.3 [a]	30 ppm

[a] Threshold values.

In this study, Perkin-Elmer Models 305B, 503, and 603 and a Varian-Techtron Model AA6 were used. In two of three laboratories, output was obtained on a potentiometric recorder.

The preferred line source was a Perkin-Elmer electrodeless discharge lamp. However, arsenic hollow-cathode lamps may also be employed. The 1937 Å arsenic resonance line was used. Other instrument parameters were similar to those recommended by the manufacturer.

A silica electrically heated absorption cell was employed. The preferred tube was 10-cm-long by 10-mm-i.d. Chromel C resistance wire 0.7 Ω/ft was wound around the length of the tube. Windings must be close to, but not touching, each other. Asbestos insulation about 8 mm in thickness was applied over the windings. An ordinary lab variac may be used to power the furnace. The tube must be heated to between 700 and 800°C for effective atomization of arsenic.

An aluminum block 7.6-cm-thick with holes 2.54 cm in diameter and 6.2-cm-deep was used to hold tubes for acid decomposition. Pyrex No. 7900 calibrated folin digestion tubes of 50 ml capacity, available from O. H. Johns Scientific, were used.

The silica tube was fastened to the burner mount of the AAS unit (e.g., Fig. 31). This arrangement simplifies the alignment of the tube in the optical beam.

Nitrogen and hydrogen gases were passed through the tube at the preferred flow rates of 225 and 75 ml/min, respectively. Nitrogen is used as a carrier gas for arsine. Hydrogen is necessary to prevent sudden wide variations in conditions inside the tube during the injection of the hydrogen/arsine mixture. The hydrogen gas burns quietly at the ends of the heated tube.

A 2% (w/v) solution of sodium borohydride was prepared by dissolving the appropriate weight in distilled water containing one pellet of

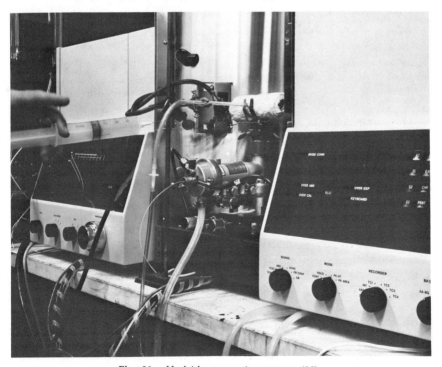

Fig. 31.   Hydride generation system (98).

KOH for every 100 ml of solution. The cloudy liquid was pressure-filtered through 0.45 μm porosity membranes. It is important to purchase sodium borohydride supplied in glass containers (e.g., Fisher certified). Sodium borohydride supplied in metal containers was often contaminated in time.

Commercially available hydrochloric acid has varying amounts of arsenic contamination. During this study, Fisher ACS reagent grade was found to be best.

Arsenic Standard and Working Solutions

To prepare a 1000 μg/ml arsenic solution, 1.320 g of primary standard, arsenious oxide is dissolved in 25 ml of 20% (w/v) potassium hydroxide solution. The solution is then neutralized to a phenolphthalein endpoint with 20% (v/v) sulfuric acid solution. Pour the solution into a 1000 ml flask, adding enough sulfuric acid to give a final concentration of 1% (w/v). Fill the flask to the mark with distilled water. To produce a stable stock solution As(III) must be oxidized to As(V). To do this, 100 ml of the 100 ppm As(III) solution is placed in a 1 liter flask; 10 ml nitric acid, 2 ml sulfuric acid, and 2 g potassium persulfate are added. The sample is digested to white sulfur trioxide fumes and then diluted to 1 liter with water. This stock solution of 100 ppm As(V) is stable for at least three months.

Working solutions of 1, 2, 5, 10, 20, 50, 75, and 100 μg/liter are prepared to contain 10% hydrochloric acid. The working solutions are stable for at least one month.

Procedure: Sample Preparation

Fusion.   This method should be used for samples containing appreciable siliceous material when total arsenic is required. The procedure is adapted from a method described by Lynch and Mihailou (99).

Place a sample (0.1–0.2 g) in a 30 ml nickel crucible. If the sample contains appreciable organic matter, an equal weight of magnesium oxide is added. Add one or two drops of ethanol. Slurry the mixture with several drops of water and swirl to homogenize. Add about 10 potassium hydroxide pellets (~1 g) for each 0.1 g of sample. Dry the contents of the crucible in an oven at 100°C. Fuse the samples in a muffle furnace at 550°C for 15 min. After cooling, leach the fusion cake with several small volumes of water. Transfer the liquid each time to a 50-ml volumetric flask. Any solid that adheres persistently to the crucible can be removed by scrubbing with a rubber policeman. After the nickel crucible has been completely rinsed, add concentrated hydrochloric acid to the solution in the flasks to give a final acid concentration of 10%. Dilute the sample to volume and mix.

Allow sediment to settle prior to the hydride generation step. Filtration may be necessary for some persistently turbid samples. Nickel crucibles must be cleaned between runs using steel wool followed by soap and water.

Acid digestion. This method should be used for organic samples without appreciable silica or when only mineral acid extractable arsenic is required. The procedure is adapted from a method by Bishop, Taylor, and Diosady (100).

Place an appropriate amount of sample in a 50 ml digestion tube marked at the 25 and 50 ml levels. Add a few drops of ethanol to the wet sample. Place the tubes in an aluminum block and add 3 ml of concentrated sulfuric acid to each. Heat the block on a hot plate set at 350°C and run 0.5 ml aliquots of hydrogen peroxide down the inside of the tubes from time to time until the solution in the tubes clears. Cool the tubes and contents, add 5 ml hydrochloric acid, and dilute the solutions to the 50 ml mark with water.

Hydride generation procedure. Draw standard and sample solutions containing between 1 and 500 μg of arsenic in 10% hydrochloric acid into the syringe to the 5 cm^3 mark, being careful to exclude air. Dip the needle into a 2% sodium borohydride solution and withdraw the plunger in one quick motion to the 50 cm^3 mark (approximately 1.5 ml of sodium borohydride is drawn into the syringe). Invert syringe and cover the needle opening with a tissue. After shaking for 15 sec insert the needle through the tygon tubing at a point 5–8 cm from the end of the glass tube on the atomizer. Inject the arsenic, as shown in Fig. 31, into the furnace. To prevent any solution from being injected into the heated tube hold the needle in a higher position relative to the body of the syringe. A stopper placed on the ribs of the plunger is used to prevent the plunger from returning to the zero position, thereby permitting an even injection rate without concern for solution entering the furnace. The arsenic peaks are monitored on a recorder. Remove the needle from the tygon tubing and discard the spent sample–sodium borohydride mixture. The syringe can be rinsed between determinations with 10% hydrochloric acid. This step was found unnecessary for most samples.

DETERMINATION OF METALS IN ROCKS AND SOILS FOR GEOCHEMICAL EXPLORATION PURPOSES (101)

Acid-Extractable Metal

Comment on the Method

The following metals can be determined: copper, nickel, zinc, lead, chromium, manganese, cobalt, iron, cadmium, and silver. The latter element requires special conditions: its concentration must be above 1 ppm (solution), and no hydrochloric acid may be used.

Geochemical exploration analyses are meant to yield relative metal concentrations between the different samples. Only the "acid-extractable metal" (e.g., loosely bound, adsorbed, or precipitated on grain surfaces) is determined by this method.

Geochemical analyses should be inexpensive and rapid because of the large number of samples that must be processed. To this end, precision and accuracy must be to some extent sacrificed. While the method is reliable in yielding relative metal contents between samples, it must never be used to indicate total metal.

Geochemical surveys require that background levels of a metal in the area be determinable. If these are close to the detection limit then a correction for nonspecific absorption is crucial.

The percentage metal extracted by the procedure varies from element to element and depends on the same type. By experiment, it was found that the extraction efficiency varies from 60 to 95%.

Reagents and Equipment

Soft glass, 15 ml test tubes and a plastic-coated wire test tube holder capable of holding 48 tubes were used. A hot water bath is easily used for dissolution (water temperature 85–95°C).

Any conventional AAS unit is satisfactory. Analyses can be speeded up considerably if the equipment reads out directly in concentration. It is debatable whether full automation is desirable since the variety of errors that occurs under these conditions is readily distinguishable if an operator is present.

Experimental conditions are similar to those recommended by the manufacturer. Working standards should be calibrated to read out directly in micrograms/gram in the soil.

Standards are prepared from 1000 μg/ml stock solutions. A mixed standard containing as many metals as desired can be used. The acid content of the standard should approximate that of the samples.

Procedure

Weigh 0.25 g samples (air-dried) into 15 ml test tubes. Place the test tubes in wire racks (48 per rack). Add 2 ml of concentrated nitric acid and allow the reaction to proceed until violent bubbling subsides. Place the rack in a water bath and heat for 30 min at 85°C. Add 1 ml hydrochloric acid (except when silver is to be done) and heat the rack for 30 min. Cool the samples and dilute to the 10 ml mark with water. (Dilution need not be accurate and may be to a scratch on the tube or some other roughly determined mark indicating 10 ml.) Cover the tops of the tubes and invert the rack several times using a board over the top to keep tubes from falling out. Allow the sediment to settle for several hours. Using appropriate standards run the desired metals. If metal values are consistently higher than the working range secondary, less sensitive lines may be used (see Table II). Background correction may be essential for cobalt, zinc, nickel, cadmium, silver, and lead when working near the detection limit. It is the author's experience that for most work, except in the case of cobalt and silver, background correction is not necessary.

Detection Limits: Elements in Soil

The detection limits of elements in soil, expressed in micrograms/gram, are as follows: copper, 5; nickel, 10; zinc, 2; cadmium, 2; iron, 5; lead, 10; cobalt, 10; manganese, 10; silver, 10; chromium, 10.

Sulfide Nickel and Copper

Comment on the Method

Over the years there have been a number of methods that have purported to selectively attack metals in a specific form. In many cases these methods worked for a specific assemblage of minerals but then failed when the specific compound was present in a different matrix. The following method may suffer from the same problem but, in the author's opinion, it offers the best possibility of success among the offerings available. Because of the importance to the mineral industry of having a sulfide selective method, the following is included with the reservation stated above.

Hydrogen peroxide is used to selectively oxidize sulfide sulfur, thus releasing metals in sulfide minerals. A buffer, ammonium citrate, is used to control the acidity to prevent attack of other forms of metal.

Reagents and Equipment

Ammonium citrate (10%).   Dilute 100 g of ammonium citrate to 1000 ml in a volumetric flask.

Reactant.   Combine 65 ml of the 10% ammonium citrate with 35 ml of 30% hydrogen peroxide and mix well.

Standards.   Prepare these to read out directly in microgram/gram with the samples. Dilute appropriate volumes of 1000 μg/ml stock solutions with reactant solution and water to give a reactant concentration similar to samples.

Some suitable mechanical shaking machine is required.

Procedure

Weigh 0.1 g of sample into a 50 ml plastic bottle. Add 10 ml of reactant and shake the solution mechanically for 2 hr at a moderate rate. Pour the samples into 25 ml test tubes, marked at the 20 ml level. Add water up to the mark. Shake the tubes to mix the solution thoroughly. Allow the sediment to settle for several hours. Analyze samples by the acid-extractable procedure above.

Detection Limit: Elements in Soil (μg/g)

Nickel, 25; copper, 25.

DETERMINATION OF METALS IN SOILS AND RELATED SOLIDS FOR ENVIRONMENTAL PURPOSES (102)

Comment on the Method

Soils, sands, etc., require pretreatment before the sample for analysis is weighed out. There is a variety of opinion on what should be done and standard methods have been proposed by various groups. Experience has shown that the following should be considered.

Samples should be sieved.   Research in the author's laboratory and elsewhere suggests that, in most cases, metal tends to associate with the fines in a soil. Hence if a sample is sieved through −200 mesh it will usually yield very much higher values than if sieving is done through −80 mesh. Each individual research requirement will determine what sieving should be done. In any case, removal of the very coarse material is essential.

Drying must be done. A good temperature for all metals except mercury is 100°C. When samples are to be done for mercury, experience suggests that drying at 75°C is permissible for most soils.

Samples should be homogenized. This can be done by rolling at least 100 g dried samples on a piece of cellophane and lifting the corners one after the other. The sample is then spread evenly over the cellophane and the sample for analysis obtained by taking random scoops from the whole sample area.

Several decomposition procedures are available depending on the requirements of the result. The following are commonly done:

(1) Total decomposition using hydrofluoric with nitric, sulfuric, perchloric, and hydrochloric acids in various combinations.

(2) Strong acid-extractable metals using aqua regia, hydrochloric, nitric, sulfuric, perchloric, or a combination of acids.

(3) "Available" metal, which involves an extraction using a weak acid such as acetic or citric, or chelating agents such as EDTA or NTA.

Reagents and Equipment

Same as procedure on p. 113.

Procedure: Total Decomposition

Weigh 1 g of sample powder into a Teflon dish. Add 25 ml hydrofluoric and 2 ml concentrated sulfuric acids. Evaporate over medium heat until fumes appear. Cool and slowly add 7 ml nitric and 21 ml hydrochloric acids. Evaporate over low heat. When dry, remove excess hydrochloric acids by adding nitric acids until brown fuming stops. Add about 15 ml water. Filter into 25 ml flask and dilute with washings to mark. Analyze directly by flame atomic absorption using instrument paramenters recommended by the manufacturer. Prepare further dilutions as required. Run standard calibration solutions and standard reference samples such as USGS standard rocks. Correction for nonspecific absorption is essential.

Procedure: Strong Acid-Leachable Metal

Use procedure on p. 134.

Procedure: Acetic Acid Extractable

Weigh 0.5–1.0 g of a dried sample into a 150 ml beaker. Add 10 ml of 0.5 N acetic acid. Digest on medium heat of a hot plate until almost dry. Add 2 ml of 0.5 N acetic acid to dissolve residue. Filter into a 25 ml flask using Whatman No. 44 paper and dilute with water. Run on flame atomic absorption spectrometer using aqueous standards with 2 ml of 0.5 N acetic acid in every 25 ml of standard. The manufacturers recommended instrument parameters are used.

DETERMINATION OF THE NOBLE METALS

Lead Fire Assay (103)

Despite frequent claims to the contrary, a fire assay is still essential for the reliable analysis of noble metals in most geological samples. The classical lead assay remains the most generally acceptable approach. Good additional information on the classical assay can be obtained from books by Beamish (103) and Bugbee (104).

No assay procedure has been developed that is applicable to all sample types. The following method may be used for the collection of platinum, palladium, gold, rhodium, iridium, osmium, and ruthenium from a wide range of materials. If osmium and ruthenium are to be analyzed, the cupellation step must be avoided. For rhodium and iridium a cupellation into gold, rather than silver, bead must be done.

Reagents and Equipment

A furnace capable of reaching 1200°C is required. Cupels roasting dishes and assay pots are available from many brick manufacturers.

Flux components must be free from noble metals but otherwise do not need to be of a high grade. Suggested flux compositions are given in Table XXIII.

Procedure: Assay

If the ore contains sulfides, the following roasting procedure must be done. Transfer the weighed ore to a 6 in. porcelain evaporating dish and place for a few minutes at the front of the furnace with the door open. The initial temperature should be about 600°C. Over a period of about 5 min move the dish to the furnace center, stirring intermittently. Partially close the door and allow to roast for 2 hr at 950°C, frequently stirring for the first 30 min to avoid agglomeration of the concentrate. Remove the dish, cool, transfer the contents to a mortar, and grind to a fine powder. Take care to avoid any loss of powder.

From the available analytical data calculate a suitable flux composition. In general, a bisilicate slag is a suitable medium for the collection of platinum metals in oxidizing ores. For samples with high proportions of associated base metals—copper, nickel, etc.—it may be desirable to increase the proportion of litharge, e.g., flux C. Decide on a

Table XXIII

Suggested Flux Compositions

| | Gold ores, | Nickel ores | | |
	Flux A (g)	Flux B (g)	Flux C (g)	Flux D (g)
Ore[a]	15	30	15	30
PbO	85.0	275	250	266
SiO	10	15	10	60
Na_2CO_3	21.1	40	12	46
CaO	4.5	12	—	12
$Na_2B_4O_7$	—	—	—	—
KNO	—	16	9	18
Flour	1–3	—	—	—

[a] For blanks, 15 g of silica replaced 15 g of gold ore and a mixture of 12 g of silica, 3 g of copper sulfide, 3 g of nickel sulfide, and 12 g of iron sulfide replaced 30 g of nickel ore.

proper ratio of flux to ore. If a silver bead is to be prepared, add to the flux an amount of silver powder or silver in solution to give a ratio of approximately 20:1 of silver to total platinum metals expected. Some suggested fluxes are given in Table XXIII.

Arrange a cellophane sheet in a suitable position, and pass the roasted ore through a No. 45 standard sieve to the center of the sheet. Add a portion of the flux to the sieve to remove any traces of ore. Transfer most of the remaining flux to the mixing sheet and mix thoroughly. Place the mixture in an assay pot of an appropriate size. Add the remaining flux to the sheet, mix, and transfer to the pot, taking care to brush the sheet free of the mixture.

Place the pots in the furnace at about 950°C, and raise the temperature at the maximum rate to 1200°C. This fusion period should be approximately 1 hr. Remove the pots, pour the mixture into conical iron molds, and allow to cool. Remove the button, taking care to retain all of the slag. Free the button of slag by gently tapping with a small iron rod. Set the button aside. Transfer the slag to a grinding mill or to a mortar, grind to pass a No. 45 standard sieve, and place the sample on the original mixing sheet. Clean the mill mortar and screen with sufficient litharge to produce a second button, then transfer to the slag on the sheet. Mix well as before and transfer the mixture to the original pot. Fuse as before and clean the lead button as described above. If necessary a fusion of the second slag can be made to produce a third button.

When the noble metals are to be concentrated to form a silver bead clean the button thoroughly by gentle tapping with an iron rod and transfer to a bone ash cupel that has been preheated at 900°C for at least

10 min. Continue heating the cupelling button at about 1000°C, with a plentiful supply of air. Remove the cupel over a period of a few minutes after the completion of the cupellation process. The cupellation bead may be parted and analyzed as given below for platinum, palladium, and gold. Lead buttons must be used for the analysis of osmium and ruthenium.

Distillation for the Recovery of Osmium and Ruthenium (103)

Comment on the Method

Cupellation is negated when osmium and ruthenium are to be analyzed. The following is the accepted procedure for the recovery of these elements from a lead button. Subsequently, atomic absorption analysis can be done on the receiver solutions.

Procedure

Transfer a lead alloy (20–50 g), obtained before cupellation by the above procedure, to the distillation flask (Fig. 32). Add 100 ml of water to the trap, and 25 ml of a 3% hydrogen peroxide solution to each of the other two receivers. Chill the receivers in an ice bath, pass water through the condenser, and apply suction to produce 2 or 3 bubbles/sec. Add 75 ml of 72% perchloric acid to the distillation flask and heat very gently until the lead is completely dissolved and effervescence of hydrogen has ceased. Continue the heating until the white fumes of

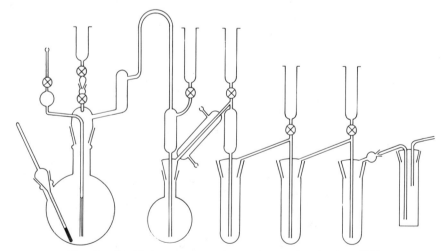

Fig. 32.   Distillation apparatus for osmium and ruthenium (103).

perchloric acid have disappeared and a colorless liquid is reflexing on the still wall.

Cool to about 60°C and add 8 ml of 36% perchloric acid. Heat again to the removal of brown fumes. Repeat the addition of 8 ml of 36% perchloric acid, intermittently heating twice to ensure the complete removal of the osmium and ruthenium tetroxides. A complete distillation requires 0.5–1 hr. Add 15 ml of 72% perchloric acid to the trap, and boil the solution for 30 min. Transfer the chilled receiving solution to a second chilled distillation flask as quickly as possible to prevent loss of osmium by volatilization, and wash the receivers thoroughly with water to remove the sulfuric acid.

Add 100 ml of water to the trap and 30 ml of a 5% thiourea solution in 1:1 hydrochloric acid to the first receiver and 10 ml of the same solution to each of the other receivers. Chill the receivers in an ice bath. Add 40 ml of a 30% hydrogen peroxide solution and 50 ml of concentrated sulfuric acid to the chilled distillation flask containing the osmium and the ruthenium distillates. Boil gently for 30 min. Add 15 ml of perchloric acid to the trap and boil for 25 min. Draw a stream of air slowly through the system (about 3 to 5 bubbles/sec). Transfer the contents of the receivers to a volumetric flask. Rinse and dilute contents to volume.

To recover the ruthenium in the pot liquid, add 100 ml of water to the trap, 30 ml of a 3% hydrogen peroxide solution, and 1 ml of 48% hydrobromic acid to the first receiver, and 10 ml of a 3% hydrogen peroxide solution to each of the other two receivers. Cool the receivers in an ice bath. Add 10 ml of concentrated sulfuric acid to the distillation flask and then add cautiously an excess of a 10% sodium bromate solution (about 20 ml). Apply a gentle suction, and distill cautiously over a low flame for 1 hr. Then add 15 ml of the perchloric acid to the trap, and boil for 25 min. Disconnect the receivers from the water condenser, maintaining the connection between the two receivers. Add 8 ml of 48% hydrobromic acid to the first receiver and 4 ml of the acid to the second receiver. Boil the liquid for 10 min. Transfer the contents of the receiver to a volumetric flask, rinse the tubes and the receivers with 10% hydrobromic acid, and dilute to volume.

Determine osmium and ruthenium using the procedures on pp. 146 and 149, respectively.

Determination of Gold, Platinum, and Palladium in Silver Assay Beads (105)

Comment on the Method

The method, as originally published, recommends sample solutions containing 6 N hydrochloric acid to keep the silver in solution. This high acid content may not be desirable when used with some equipment because of its corrosive properties. It is possible to work in dilute hydrochloric acid solutions as long as the silver chloride is allowed to settle prior to the analysis.

A variety of interference problems were encountered. These were eliminated by using 1% lanthanum as a releasing agent.

Reagents and Equipment

A Perkin-Elmer 303 atomic absorption spectrophotometer fitted with a Boling burner was used. The following wavelengths, slit widths, and lamp currents were used, respectively: platinum, 2659 Å, 10 Å, 25 mA; palladium, 2476 Å, 10 Å, 30 mA; gold, 2428 Å, 30 Å, 14 mA.

Standard solutions were prepared from pure metals of each material. Other reagents were Fisher analyzed grade and were checked for the absence of the precious metals.

Procedure

Add 5 ml of nitric acid to a silver bead in a 50 ml beaker and heat to leach the silver. Evaporate the solution to 0.5 ml and add several milliliters of concentrated hydrochloric acid. After all bubbling has ceased add additional portions of concentrated hydrochloric acid until no further gases are evolved. Add enough lanthanum to make the final concentration in the flask 1%. Wash the mixture into a volumetric flask and dilute to volume with 6 N hydrochloric acid. Determine the platinum, palladium, and gold absorbances against standards containing 1% lanthanum and 6 N hydrochloric acid by running closely a lower concentration standard, the sample, and higher concentration sample in quick succession.

Aspirate water frequently to minimize corrosion.

Detection Limit. In these sample solutions, the detection limits (μg/ml) are palladium, 1; gold, 1; platinum, 5.

Determination of Iridium (52)

Comment on the Method

Cupellation into silver assay beads is not permissible because of the insolubility of iridium in silver. The following procedure involves a classical fire assay followed by cupellation into a gold bead. A copper–sodium sulfate buffer system is employed to overcome any potential interferences. The procedure was found to give excellent results for iridium on a standard dunite sample. The fire assay is also suitable for rhodium, platinum, and palladium, although no details are given by the authors for the atomic absorption determination of these elements.

Reagents and Equipment

Gold wire, 99.999% pure, was used. For the mixed copper–sodium solution, dissolve 13.75 g of copper sulfate pentahydrate and 4.64 g of sodium sulfate in 50 ml of hydrochloric acid and make up to 100 ml with water.

Standard solutions of iridium. Prepare from ammonium hexachloroiridate a stock solution containing 1.000 mg/ml iridium in 2% v/v hydrochloric acid. Prepare other solutions by dilution with 2% v/v hydrochloric acid.

Cylindrical alumina crucibles, 2 ml capacity, were used for sodium peroxide fusions. Coors AD-999 alumina ceramic crucibles are available from Coors Porcelain Company, Golden, Colorado.

A Perkin-Elmer Intensitron hollow-cathode tube with a Perkin-Elmer Model 303 instrument was used in the experiments. The conditions were as follows: wavelength, 2640 Å; slit, 3 Å; hollow-cathode current, 30 mA; fuel–acetylene pressure, 69 kN/m^2 (10 psi); flowmeter setting, 8.5; oxidizer–air pressure, 190 kN/m^2 (28 psi); flowmeter setting, 7.5; flame, oxidizing; burner, standard head; aspirator, adjusted for uptake of 3 ml/min. Conditions for other instruments must be found by trial and error.

Procedure

The fire assay fusion and cupellation procedures used here follow generally accepted practices as described by Bugbee (104). A bisilicate slag composition is to be preferred. For a 20 g sample of dunite, a flux consisting of 50 g of lead(II) oxide, 35 g of sodium carbonate, 159 g of silica, 19 g of sodium tetraborate, and 4 g of flour will yield both a satisfactory lead button and a bisilicate slag. The fusion is made in the presence of 50 mg of added gold for the quantitative collection of iridium, palladium, platinum, and rhodium.

Transfer the gold bead obtained on cupellation to a small beaker. Add 5 ml of aqua regia, cover, and allow the mixture to react at room temperature for approximately 1 hr. Heat on a steam bath for several hours more. Add 3 ml of water and a small amount of paper pulp; mix, and filter through a 42.5 mm medium porosity filter paper. Wash the iridium residue with water. Ignite the residue in a 2 ml alumina crucible at 600°C starting with a cold furnace. Add 100 ± 10 mg of sodium peroxide by calibrated dipper and carefully fuse the residue. Heat for 5 min more, maintaining in the molten state, and then cool. Add 1.5 ml of water, cover, and allow to stand at room temperature for approximately 15 min. Warm the mixture until the melt disintegrates, and transfer the solution to a 25 ml beaker by several alternating washes with 1 ml portions of concentrated hydrochloric acid, nitric acid, and water. Add 0.15 ml of sulfuric acid 1:1 by pipet to convert sodium salts into sulfate and evaporate the solution on a steam bath. Add 2 ml of hydrochloric acid (1:1), warm briefly to dissolve salts, and transfer the solution to a 10 ml volumetric flask with water. Dilute to volume with water and mix. Transfer a 5 ml aliquot of the solution to another 10 ml volumetric flask. Add 1 ml of hydrochloric acid (1:1) and 1 ml of copper sulfate solution. Adjust to volume with water and mix. Prepare iridium standards and a blank, each containing 1 ml of mixed copper–sodium solution in a 5-ml volume. Determine iridium in all solutions by atomic absorption. Samples containing as low as 2.5 ppm iridium can be determined.

Determination of Rhodium in Chromite Concentrates (51)

Comment on the Method

Two procedures for the collection of rhodium are given, a tellurium precipitation and a fire assay. In the case of the former, a sodium peroxide fusion of the sample is followed by dissolution and then coprecipitation of the rhodium with tellurium. The fire assay procedure includes a cupellation step yielding a gold bead. Atomic absorption analysis can be performed on sample solutions of either product. To overcome potential interferences, 1% lanthanum sulfate is used.

Reagents and Equipment

Tellurium solution. Prepare 1 mg/ml in 10% HCl. Dissolve tellurium metal in aqua regia and remove nitrate by evaporation with hydrochloric acid.

Tin(II) chloride solution. Dissolve 20 g of fresh stannous chloride dihydrate in 17 ml of hydrochloric acid. Dilute to 100 ml with water.

Gold wire for fire assay. Wire 99.999% pure and 0.1 mm in diameter. Cut into 2.5 mg segments.

Lanthanum sulfate solution. Dissolve 14.66 g of lanthanum oxide in 25 ml of hydrochloric acid and then add 15 ml of 1:1 sulfuric acid.

Evaporate the solution. Dissolve the residue in 125 ml of hydrochloric acid and dilute to 500 ml with water.

Standard solutions of rhodium. Prepare from the ammonium chloro-salt a stock solution containing 1.00 mg/ml rhodium in 2% v/v hydrochloric acid. Prepare other solutions by dilution by factors of 10 with 2% v/v hydrochloric acid.

Alumina crucibles for sodium peroxide fusions. Coors AD-999 alumina ceramic, available from Coors Porcelain Company, Golden, Colorado.

Instrument parameters and settings. A Perkin-Elmer Model 303 instrument was used with the following operating conditions:

Wavelength	3435Å
Slit	0.3 nm
Hollow-cathode current	20 mA
Acetylene flow setting	6
Air flow setting	6.8
Flame	Oxidizing
Burner	Standard head
Aspirator	Adjusted for optimum uptake

Procedure

Tellurium precipitation procedure. Fuse over a burner in an alumina crucible 3.0 g of chromite concentrate with 10 g of fresh sodium peroxide. Heat for approximately 15 min after the charge becomes molten. After cooling the melt, place the crucible in approximately 100 ml of water in a beaker and carefully add 60 ml of hydrochloric acid. Detach the melt and remove the crucible.

Heat the solution to approximately 60°C and then, while stirring, very carefully add 5–7 ml of 30% hydrogen peroxide. Heat this solution on the steam bath for 30 min or more to destroy peroxide and then filter the solution through Schleicher and Schüll 589 White Ribbon paper (or equivalent).

Add 2.5 ml of tellurium solution, then 25 ml of tin(II) chloride solution by pipet while stirring. Adjust solution volume to approximately 200 ml. Digest the tellurium precipitate on a steam bath for approximately 2 hr, filter it off on a Schleicher and Schüll 589 White Ribbon paper, and then wash with hot 10% v/v hydrochloric acid. Discard the filtrate.

Dissolve the precipitate off the paper by slowly adding 50 ml of hot aqua regia (8 volumes of hydrochloric acid, 2 of nitric acid, and 5 of water), collecting the filtrate in a 100 ml beaker. Wash finally with hot 10% v/v hydrochloric acid. Pass this filtrate through a 15 ml medium-porosity fritted glass Buchner-type filter funnel to remove paper fibers. Wash the filter with 10% v/v hydrochloric acid and then evaporate, treating with 3 ml portions of hydrochloric acid.

Add 2 ml of lanthanum sulfate solution to the residue and warm briefly on the steambath. Transfer the solution to a 5 ml volumetric flask and adjust to volume with water. Prepare ruthenium standards and a blank containing 2 ml of the lanthanum solution in a 5 ml volume. Determine rhodium on all solutions by atomic absorption.

Fire assay procedure. Add 3.0 g of sample to a flux consisting of 35 g of sodium carbonate, 11 g of silica, 19 g of anhydrous sodium tetracarborate, 50 g of lead oxide, and 4.2 g of flour contained in a fire assay crucible and mix thoroughly.

Place in a furnace at 850°C and gradually raise the temperature to 925°C. Heat for 10 min at this temperature. Total heating time should be approximately 50 min. Pour the melt into an iron mold. Collect the lead button and shape it into a cube. Make a linear indentation on one surface of the cube by tapping a knife edge against this surface. Place a segment of the gold wire in the indentation and then carefully hammer the cube to secure the wire in place. Cupel the lead button at approximately 950°C. Analyze the bead as for iridium.

Determination of Osmium in Solutions (106)

Comment on the Method

This procedure is applicable to osmium in aqueous thiourea and chloroform solutions. Although not tested for this purpose, the procedure is most likely useful in work with receiver solutions after the distillation separation of osmium as the tetroxide. This would seem best accomplished by choosing the thiourea receiver solution.

Reagents and Equipment

A Jarrel-Ash atomic absorption spectrometer model 82-526 with 3000 Å grating, type R-106 photomultiplier tube, 5 cm laminar flow burner (Aztec Instruments, Inc.), and Westinghouse osmium hollow-cathode lamp was used.

Osmium tetroxide was obtained from Mallinckrodt in 1 g ampules.

Use the primary osmium resonance line of 2909 Å. The experimental parameters established are 8 mA current to the cathode, 12 psi acetylene, 32 psi nitrous oxide, 18.25 mm burner height, and either 700 or 720 V applied to the photomultiplier. The reddish part of the flame is about 0.5 in. in height. Adjust the burner height so that the light emitted from the hollow cathode tube will pass directly over the thin white part of the flame, since the concentration of neutral dissociated atoms is highest in this region of the flame.

Standard osmium solutions. Prepare standards by dissolving osmium tetroxide from the ampules in chloroform or water. These may be standardized by treatment with thiourea, or by precipitation with thionalide. Prepare working solutions by dilution of the above.

Procedure

Determination of osmium in water or chloroform solutions. Adjust the equipment as indicated in the instrumental parameters and aspirate the solutions directly. Owing to incomplete combustion of chloroform, the burner slot must be cleaned after each analysis. The glass burner chamber and burner head are cleaned periodically of any solid residue on the inside surfaces. The combustion products must be vented into a hood.

Determination of osmium in thiourea complexes. The thiourea complex is a familiar occurrence for Os during analytical manipulations. Before the atomic absorption determination can be performed the thiourea complex must be destroyed as follows: add 5 ml of a solution of the complex to a 25 ml volumetric flask with 2 ml of 30% hydrogen peroxide. Dilute each flask to the mark with distilled water, stopper and allow to stand at room temperature for 2.5 hr for the reaction to proceed to completion. Then determine osmium in the flask by direct aspiration.

Determination of Ruthenium and Osmium in Leaching Residues (107)

Comment on the Method

A distillation separation of osmium and ruthenium from complex leach residue solutions is employed. Osmium is determined colorimetrically using thiourea and ruthenium is analyzed by atomic absorption. The colorimetric procedure rather than atomic absorption is preferable for osmium. Uranium can be used to suppress any interferences in the atomic absorption determination of ruthenium.

Reagents and Equipment

The distillation apparatus is shown in Fig. 33. Barium peroxide A.R. grade was used. Hydrochloric and perchloric acids were 6 N and 70%, respectively.

Uranium solution.   Dissolve 59 g of pure uranium oxide as U_3O_8 in 20 ml of aqua regia. Filter, if necessary, and dilute to 200 ml with water (1 ml is equivalent to 250 mg of uranium).

Hydroxylamine hydrochloride, 10% (w/v).   Dissolve 10 g of hydroxylamine hydrochloride in 100 ml of distilled water.

Thiourea solution, 10% (w/v).   Dissolve 255 g of thiourea in water, and dilute to 25 ml with water.

Stannous chloride solution, 10% (w/v).   Dissolve 2.5 g of stanous chloride dihydrate reagent grade in 8 ml of hydrochloric acid. Warm gently to dissolve. Cool and dilute to 25 ml with water.

Procedure

Sample procedure and distillation of osmium and ruthenium.   Transfer an appropriate amount of sample (see Table XXIV) to a

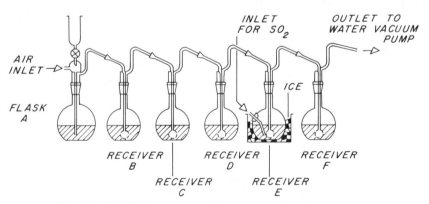

Fig. 33.   Distillation apparatus for osmium and ruthenium (107).

Table XXIV

Mass of Sample and Dilution Required

Estimated Os (μg)	Mass of sample (g)	Dilution of distillate (ml)	Aliquot portion (ml)
0–300	1.0	50	20
300–720	0.5	50	20

25 ml nickel crucible. Add 20 g of barium peroxide per gram of sample, and mix well by stirring with a thin plastic rod.

Place the nickel crucible in a muffle furnace at 700°C for 16 hr (at this temperature the mixture will not fuse but will form a sinter). Cool the nickel crucible. To remove the sinter, invert the crucible onto glazed paper. Referring to Fig. 33 transfer the sinter to a clean, dry, 500 ml distillation flask (A). Place 50 ml of 6 N hydrochloric acid in each of five 250 ml receivers, and connect the distillation apparatus. Cool receiver E in ice. Using a water vacuum pump, draw sulfur dioxide through receivers E and F, and air through the complete distillation train for 10–15 min (note 1) at an airflow between 0.2 and 0.5 ft³/hr. Increase the air flow to full pressure, and slowly add 75 ml of concentrated perchloric acid (note 2). When the reaction between the perchloric acid and the sinter has ceased, decrease the airflow to between 0.2 and 0.5 ft³/hr. Gently heat the contents of the flask with a burner until the sinter has disintegrated. Increase to the full heat of the burner, and boil the contents of the flask for 15 min. Remove the heat source from the distillation flask, and bring receiver B to the boil, using a gas burner at a low heat. Disconnect the distillation flask from the distillation train, and allow receiver B to boil for 5 min. Bring receiver C to the boil, remove heat source below receiver B, and allow receiver C to boil for 5 min. Bring receiver D to the boil, remove heat source below receiver C, and allow receiver D to boil for 7 min (note 3). Remove the heat source, and disconnect the distillation train by removing first receiver B and then the others in sequence. Combine the contents of receivers B–D in a 400 ml beaker for the determination of ruthenium. Combine the contents of receivers E and F in a 500 ml beaker for the determination of osmium.

Determination of ruthenium. Evaporate the ruthenium distillate until the volume is reduced to about 30 ml (note 4). Cool and dilute to 50 ml with water. Transfer an aliquot portion containing 100–500 μg of ruthenium to a 10 ml volumetric flask and add 0.4 ml of the uranium solution and 0.8 ml of aqua regia (note 5). Dilute to the mark with 40% hydrochloric acid. If an aliquot portion of more than 3 ml is required, evaporate the respective aliquot portion to about 3 ml, add 0.4 ml of uranium solution and 0.8 ml of aqua regia, and after transferring the solution to a 10 ml volumetric flask, dilute to volume with 40% hydrochloric acid.

Using the following instrumental parameters on the atomic absorption spectrometer, measure the absorption of the ruthenium and compare it with that of suitable standards measured at the same time:

Wavelength	3499Å
Width of slit	50 μm
Lamp current	10 mA
Air pressure	15 lb/in.^{2}

Acetylene flow is adjusted to give optimum readings. In the same way, measure the absorption of a reagent blank and subtract the value obtained from the values obtained from the samples.

Prepare a 2000 μg/ml standard ruthenium solution from metal or sponge. Dilute this solution so that it contains 100 μg Ru/ml in 30% hydrochloric acid. Take aliquot portions of 1, 2, 3, 4, and 5 ml, add 0.4 ml of uranium solution, 0.8 ml of aqua regia, and dilute to 10 ml in a volumetric flask with 40% hydrochloric acid. This procedure gives standard ruthenium solutions of 10, 20, 30, 40, and 50 μg Ru/ml. Measure the absorption of standards and samples relative to a blank solution containing uranium, aqua regia, and 40% (v/v) hydrochloric acid in a 10 ml volume.

NOTES

1. Before the distillation, the 6 N hydrochloric acid in receivers E and F must be saturated with sulfur dioxide so that the osmium tetroxide is reduced to osmium sulfite.
2. The perchloric acid oxidizes the ruthenium and the osmium to the volatile octavalent state.
3. The successive boiling of receivers B–D transfers the osmium into receivers E and F where it is reduced to osmium sulfite by the sulfur dioxide. The ruthenium remains in receivers B–D.
4. Because the perchloric acid fumes condense in the various receivers, it is not advisable to evaporate the ruthenium distillate to a volume less than 30 ml. The perchloric acid will oxidize the ruthenium to the volatile octavalent state.
5. The routine atomic absorption analysis for ruthenium is usually done in the presence of other platinum group metals and gold, and uranium is added for the elimination of interference effects. In the determination of ruthenium given here, uranium is added to the solutions for analysis so that standards and samples can be matched.

Solvent Extraction Method for Gold (108)

Comments of the Method

For this very simple method to be applicable, the sample must yield its gold to aqua regia dissolution. The worker is cautioned that aqua regia does not always extract gold quantitatively from complex rock matrix. If the gold is present as the metal and if all the finely disseminated blebs are exposed on the grain surfaces by grinding them, aqua regia attack can quantitatively dissolve gold. If any uncertainty exists regarding this point a lead fire assay approach should be used.

Reagents and Equipment

A Jarrell Ash 810 atomic absorption spectrometer with a Westinghouse gold hollow cathode was used for the atomic absorption measurements. The monochromator was set at 2428 Å and instrument parameters were applied as recommended by the manufacturer. Reagent grade acids and methylisobutyl ketone (MIBK) were used.

Preparation of standards. Weigh accurately 100 mg of gold and place into a 100 ml volume flask. Add 5 ml of aqua regia to dissolve the gold. Dilute to mark with distilled water.

From this stock solution, prepare 100 ml portions of 10, 50, 100, 500, 1000, and 5000 ppb gold by appropriate dilution, always maintaining a 5% aqua regia acid level in all standards. Transfer the standards into 150 ml Erlenmeyer flasks and proceed with the extraction as outlined below.

Procedure

Samples high in graphite or sulfide must be preroasted. Place 5.0 g of finely ground sample pulp (95% –200 mesh) into a 200 ml Erlenmeyer flask. Add 20 ml of aqua regia. Shake the sample mechanically for 60 min. Filter the solution using No. 40 Whatman paper. Wash twice with distilled water and discard the precipitate. Dilute the filtrate to ~100 ml. Add 15 ml of MIBK, cap the flask, and shake vigorously for 2 min. If MIBK layer is milky, add a few drops of hydrofluoric acid. Aspirate the MIBK layer for the determination of gold.

Determination of Platinum, Palladium, Rhodium, Iridium, and Ruthenium after Solvent Extraction (109)

Comment on the Method

The five noble metals are separated from the base metals iron, copper, nickel, and cobalt by extraction with N-octylaniline. Lanthanum and neodymium nitrates are used as releasing agents. Using the organic layer, an enhancement of 2 to 4 is obtained for rhodium, platinum, and palladium, in an air–acetylene or nitrous oxide–acetylene flame and for ruthenium in a nitrous oxide–acetylene flame. With an air–acetylene flame the sensitivity for ruthenium and iridium is best when the aqueous solution is used.

This extraction procedure has the distinct advantage of working in acid solution. Under these conditions the technique may be readily applicable to acid digests of samples.

The procedure was tested on nickel powder, copper, nickel solution, copper slime, anode nickel, and nickel slime and the results compared favorably with those obtained by another method.

Reagents and Equipment

Solutions of metals. $H_2PtCl_6 \cdot 6H_2O$, $PdCl_2$, $RhCl_3 \cdot 4H_2O$, K_2IrCl_6, and $K_2Ru(H_2O)Cl_5$ were used as standard compounds. They were dissolved in hydrochloric acid of appropriate concentration. Rhodium chloride was first prepared in 11 M hydrochloric acid.

Extraction solution. A 2 M solution of octylaniline in toluene was shaken three times with equal volume of 3 M hydrochloric acid for 3 min each time.

Atomic absorption spectrometer. A Perkin-Elmer Model 4A was used. The flames were air–acetylene and nitrous oxide–acetylene. The analytical lines were platinum, 2659 Å; palladium, 3404 Å; rhodium, 3435 Å; ruthenium, 3499 Å.

The solutions were introduced into the flame under the following conditions: For aqueous solution and the air–acetylene flame, the flow rates (liters/min) were air, 8; auxiliary air, 10.3; acetylene, 2.2 (for iridium and ruthenium). For organic solutions the flow rates were air, 12.7; auxiliary air, 11.9; acetylene, 2.2. For the nitrous oxide–acetylene flame, the flow rates were nitrous oxide, 13.5; air, 7.5 (aqueous solutions), and 5.2 (organic solutions).

Procedure

The platinum metals are extracted from 3 M hydrochloric acid solutions with three portions of octylaniline solution (extraction time 15 min). The combined organic phases are scrubbed twice with equal volumes of 3 M hydrochloric acid (5 min shaking) to remove copper, cobalt, nickel, and iron. In order to avoid loss of palladium and rhodium to the wash solution, the latter is extracted once with octylaniline solution and all the organic phases are combined.

The extract obtained is evaporated in a corundum crucible to remove the solvent and some of the extractant, and the residue is ignited at 650°C. The residue is fused with sodium peroxide and the resulting glass is dissolved in 6 M hydrochloric acid. The solution is filtered, the filtrate evaporated, and the residue dissolved in 2 M hydrochloric acid. To determine platinum, rhodium, iridium, and ruthenium enough lanthanum or neodymium nitrate is added to give a 1% concentration of the metal. The same concentrations of lanthanium or neodymium are used in the standard solutions of the platinum metals. The atomic absorption is measured in the air–acetylene flame. To determine only platinum and rhodium, lanthanum chloride is used as the buffer. The content of palladium is found with standard solutions containing no additive. The coefficient of variation is 4–6.8%.

To determine platinum, palladium, iridium, rhodium, and ruthenium directly in the octylaniline phase, enough lanthanium solution is added to give a 1% lanthanum concentration, and the atomic absorption is measured: that of palladium, platinum, rhodium, and ruthenium in the air–acetylene flame, that of iridium in the nitrous oxide–acetylene flame. The buffer is also added to the standard solutions. For palladium, standard solutions containing no buffer may also be used. The coefficient of variation is 2–6.8%. Detection limits are given in Table XXV.

Rapid-Fire Assay Method for the Determination of Platinum, Palladium, and Gold in Ores and Concentrates (110)

Comment on the Method

A modification of the tin collection method is used to preconcentrate the noble metals platinum, palladium, and gold in ores and concentrates. Tellurium is employed to prevent dissolution of these metals during parting with hydrochloric acid. For this purpose, 15–20 mg of tellurium is sufficient. Although tellurium causes a slight depression in the absorbances of the noble metals it presents no problems at the concentrations used.

The procedure was verified for the determination of platinum, palladium, and gold in four certified reference materials prepared by the Canadian Reference Materials Project and using a standard South African ore.

Table XXV

Detection Limits of Platinum Metals (μg/ml)

	Air–acetylene flame		Nitrous oxide–acetylene flame	
	2 M HCl	N-octylaniline	2 M HCl	N-octylaniline
Pt	1.25	0.51	3.75	1.28
Pd	0.16	0.04	1.12	0.15
Rh	0.07	0.02	0.24	0.05
Ir	5.00	3.50	10.65	10.55
Ru	1.65	3.00	1.80	0.62

Apparatus and Reagents

Furnaces. A 15 kW Globar type with suitable thermocouple and temperature controller, capable of accommodating six 40 g assay crucibles and maintaining their temperature at 1250°; a Jelrus Handy-Melt portable electric furnace (Jelrus Technical Products Corp., New Hyde Park, New York); or similar. The Jelrus is a small vertical furnace equipped with removable graphite crucibles used in this work for melting tin-based assay buttons before their granulation in water.

It is recommended that after 4 to 6 months of relatively constant use the bottom of the crucibles be examined for small holes.

Tin(IV) oxide. BDH reagent grade is preferred because it has consistently given a gold blank value of 150–200 ng/g.

Tellurium powder used was reagent grade.

Standard solutions of platinum, palladium, and gold. Prepared by dissolving accurately weighed quantities of Johnson Matthey "specpure" sponge in aqua regia. Each solution is evaporated to dryness, then the residues are dissolved in concentrated hydrochloric acid, and the solution evaporated to dryness again, this being repeated several times. Finally, the salts are dissolved in, and diluted to volume with, 1 M hydrochloric acid. The gold solution is standardized gravimetrically by the classical fire assay procedure with lead, and the platinum and palladium solutions are standardized spectrometrically.

Mixed cadmium–copper sulfate solution. Prepare by dissolving 98 g of $CuSO_4 \cdot 5H_2O$ and 57 g of $3CdSO_4 \cdot 8H_2O$ in 500 ml of 12 M hydrochloric acid and 300 ml of water, followed by dilution to liter with water.

Flux for fire assay. Stannic oxide 40 g, sodium carbonate 50 g, sodium tetraborate 10 g, flour 35 g, tellurium 25 mg, silica 10–20 g according to the amount of silica in the sample. Make enough flux for a sample up to 1 assay ton (29.17 g) in size.

Procedure

Before the crucible fusion procedure, all samples except those of copper–nickel matte are roasted at 750–800° for approximately 1 hr to decompose sulfides and volatilize arsenic and antimony. The sample is placed on a shallow fire-clay dish and stirred intermittently during the roasting process. In cases where only a few grams of material (particularly sulfides) are to be roasted, the sample is placed on a bed of silica to prevent possible loss of the resultant calcine to the surface of the dish (the silica is included as part of that required in the flux above).

Leaching is performed to remove the bulk of the copper and nickel from the residue of precious metals. The sample, weighing up to 2 assay tons, is placed in a 1500 ml beaker and treated with 25 g of ammonium chloride and 100–200 ml of 12 M hydrochloric acid. The sample is heated until the amount of insoluble matter appears not to exceed 2–3 g. With large samples, it may be necessary to treat the residue once or twice more with fresh acid after intervening filtrations.

The combined sample solution (~100 ml) is diluted with an equal volume of water and the solution containing most of the nickel and copper is filtered through a moderately fast paper. The solids are completely washed onto the paper with dilute (~5%) hydrochloric acid. The washed residue and paper are dried at ~100° for about 1 hr and then mixed with the recommended assay flux for fusion.

Chromite is not completely decomposed during the fusion process, and samples containing appreciable proportions must be subjected to a pretreatment. Decomposition is accomplished by sintering with sodium peroxide. The sample is mixed with 1.5 times its weight of sodium peroxide, then placed on a 10 g bed of silica in a roasting dish, and roasted at 700° for about 1 hr. The sinter cake and underlying silica are ground together in a mortar and mixed with the flux for the crucible fusion process. The weights of sodium peroxide and silica are subtracted from the weights of sodium carbonate and silica, respectively, in the flux described above.

For powdered samples, the standard assay practice of blending the samples with the flux on glazed paper and transferring the charge to a 40 g crucible is followed.

When solutions are to be mixed with the flux, approximately one-third of the flux is placed in the crucible and a 30-cm square of thin, commercial wrapping film is pressed into the crucible to form an envelope, and then the remainder of the flux is transferred slowly in the depression so as to avoid wetting the film or crucible walls. The crucible is then heated in a drying oven at 110° for at least 2 hr. After drying, the material in the wrapping film is ground in a mortar, mixed well, and placed back in the film in the crucible. It is to be noted that, after drying, the salted portion of the charge is lumpy and difficult to pulverize and mix with the rest of the charge. This could lead to occasional spurious results.

Fusion. The crucible is placed in the assay furnace at 1250°C for about 90 min to fuse the charge. At the completion of the fusion period, the melt should not be viscous or lumpy nor should there be extensive crust formation at the top of the melt.

The melt is poured into a conical steel mold and, when it is cool, the tin button is separated from adhering slag by tapping with a small hammer.

The button is placed in the crucible of the Jelrus furnace, from which air is purged by nitrogen delivered through a ceramic tube placed directly over the button. The temperature is increased until the button melts (600–1000°, depending upon composition) and then the melt is poured into a pail of water to granulate the alloy. Any large pieces are easily reduced in size with metal shears.

Dissolution and analysis. Each sample of granulated tin alloy is treated with 150 ml of 12 M hydrochloric acid in a covered 600 ml beaker and heated until the excess of tin has dissolved and vigorous evolution of bubbles from the residue has ceased. A further 15–25 ml portion of acid is added and the sample is boiled for approximately 10 min. Water is added to give a volume of approximately 400 ml and the residue is allowed to settle. The supernatant solution is decanted through a filter pad. The residue in the beaker is washed several times, by decantation, with 15% v/v hydrochloric acid, the washings being passed through the filter pad.

The residue in the beaker is treated with a mixture of 15 ml of 12 M hydrochloric acid and 5 ml of 30% hydrogen peroxide, and the beaker is heated gently for a few minutes to ensure complete dissolution of the residue. The residue on the filter pad is eluted with 20 ml of a 3:1 mixture of 8 M hydrochloric acid and 30% hydrogen peroxide, and added to the beaker.

Approximately 50 mg of sodium chloride is added, and the sample solution is evaporated to dryness. When the evolution of fumes has nearly ceased, the beaker is removed from the evaporator and the sides are washed with ~10 ml of a 7:2 mixture of hydrochloric and hydrobromic acid. The sample is again evaporated to dryness to volatilize the remaining tin. The beaker is cooled, 10–15 ml of 12 M hydrochloric acid are added, and, while the beaker is being swirled, 30% hydrogen peroxide is cautiously added until it is evident that an excess is present. The beaker is heated for a few minutes, and, after cooling the sides are washed with water. After filtration of the solution through a fast paper into a 400 ml beaker and washing of the paper several times with 15% hydrochloric acid, approximately 5 ml of aqua regia are added and the solution is evaporated to approximately 1 ml.

To the cooled sample solution, 5 ml of cadmium–copper sulfate solution is added and the mixture is transferred to a 25 ml volumetric flask and diluted to volume with water. The platinum, palladium, and gold content of the sample is then determined by atomic absorption using the manufacturers recommended conditions. Any silver, rhodium, ruthenium, or iridium remaining in the solution will not interfere.

Note. For milligram amounts of the precious metals, the solution obtained after the volatilization of tin is filtered into a 100–500 ml flask and diluted to volume with 15% hydrochloric acid to prevent hydrolysis. An aliquot is taken and treated by the procedure given above.

Calibration curves for gold and palladium are linear in the ranges 0.2–3 and 0.4–3 ppm, respectively.

Because all batches of stannic oxide tested in this laboratory were found to contain gold, it is deemed necessary to carry a blank through the analytical scheme.

4
Analysis of Organic Samples

Organic samples span a wide range of organic chemical composition. Each sample type presents unique problems of sampling, sample preservation, storage, and physical and chemical pretreatment.

Sampling is beyond the scope of this book. The analyst must assume that the samples that have been given are representative and uncontaminated. Thiers, referring to contamination of organic samples, has stated (111): "Unless the complete history of the sample is known with certainty, the analyst is well advised not to spend his time analyzing it."

Sample preservation, storage, and physical pretreatment are also mainly the responsibility of the sample collector. Most of these steps must be done at or very soon after the sampling time. In general it is important to analyze organic samples as soon after collection as possible. Commonly used pretreatment techniques are freeze-drying, air-drying, and oven-drying (60–80°C). The method chosen depends on the sample type and the equipment available. If sample preservation is necessary, it is essential for the analyst to know the procedure employed in order that proper blanks may be run on any preservative additives. When drying is done, the problem of how to relate fresh sample weight to actual sample weight must be addressed.

Samples are generally ground to pass a 30–80 mesh sieve. Grinding is normally accomplished with a Wiley mill. Contamination during grinding and sieving can be a problem.

DECOMPOSITION OF SAMPLES—GENERAL

In general there are three approaches to the decomposition of organic matter: dry-ashing, wet-ashing, and oxidative fusion. Dry-ashing minimizes contamination, which can be an acute problem when reagents must be added as is the case in wet oxidation or oxidative fusion.

On the other hand, the relatively high volatility of some elements plagues the dry-ashing approach. Mercury is volatilized well below the ashing temperature of organic matter. In many samples, mercury is lost below the 100°C drying temperature. Special procedures are given below for this element. Cadmium is generally believed to be volatile at temperatures above 450°C, although some workers report losses below 400°C. The chemical form of the metal will influence its volatility temperature and is likely the reason for many such disagreements.

Elements such as selenium and arsenic can form volatile hydrides under reducing conditions that inevitably occur during charring. Zinc and lead are lost at temperatures in the range 300–1000°C.

Recently, low-temperature dry-ashers utilizing oxygen plasmas have been introduced. These are expensive and excessively time-consuming. More importantly, they have not been sufficiently evaluated to be adapted for routine use.

Wet-ashing is the generally accepted technique. Of the acids available, most workers recommend a combination of nitric and perchloric acids with or without other mineral acids. There is some difference of opinion on ratios of the acids that should be used. However, it is of great importance to have an excess of nitric acid present at the beginning to decompose easily oxidizable material and thus minimize chance of perchloric acid explosion. (See p. 169 for a list of precautions.)

Nitric acid is a good oxidizing agent and hence is potentially suitable for decomposition of organic matter. However, its low boiling point results in excessive loss of acid by evaporation. Nitric acid is frequently used in combination with other mineral acids when samples high in chloride are decomposed. This prevents loss of some metals as volatile chlorides.

Sulfuric acid is an oxidizing and dehydrating agent with a relatively high boiling point. Unfortunately this acid, when used alone, yields a black, charred residue. This can result in the loss of hydride-forming elements.

Some organic tissue, particularly the leaves and stems of plants, may contain appreciable silicon. Dry-ashing of these materials may result in loss of metals to the mineral acid, insoluble silica. In this case, hydrofluoric acid must be used, together with mineral acids, in the final dissolution.

Perchloric Acid Decompositions

Digestion with perchloric acid is the most widely accepted approach for the decomposition of organic samples. Martinie and Schilt (112) studied the effectiveness of perchloric–nitric acid and perchloric–nitric–sulfuric acid mixtures for the decomposition of 87 organic substances. The following methods were tested.

Nitric and Perchloric Acids

Approximately 1 g of sample was accurately weighed and transferred into a 125 ml conical flask, and 15 ml of 2:1 perchloric–nitric acid mixture was carefully added behind a protective screen. The mixture was observed for a short period for any violent behavior, then placed on low heat under the fume eradicator, and brought slowly to 120°C, at which point nitric acid began to distill. Occasionally it was necessary to interrupt the heating during the 20–120°C interval because of violent foaming. The heating rate was increased to remove the nitric acid gradually, requiring about 15 min to raise the temperature of the mixture from 120 to 140°C. Once 140°C was reached, the temperature rose rapidly to 203°C, the boiling point of the perchloric acid–water azeotropic composition (72.5% perchloric acid). At this stage, if foaming again occurred, it was sometimes necessary to interrupt the heating again. The sample was maintained at a gentle boil at 203°C for 30 min and then removed from the hotplate to cool to room temperature.

Sulfuric, Nitric, and Perchloric Acids

A 1 g sample was weighed into a 125 ml conical flask, treated with 15 ml of sulfuric acid, and heated at boiling for 15 min. After allowing the solution to cool, 15 ml of nitric acid was added, and heating was resumed at a rate to cause the nitric acid to distill over a 15 min period and the temperature to rise to 320°C before heating was interrupted. The solution was allowed to cool, 15 ml of perchloric acid was added, and heating was resumed at such a rate as to cause perchloric acid to distill from the sulfuric acid solution. The solution that remained was allowed to cool at room temperature.

Following decomposition, the residue was examined and checked for organic matter content.

The study showed that most nitrogen-containing compounds gave a residue of ammonium perchlorate on cooling. Using the perchloric acid–nitric acid method, the following compounds gave violent reactions: lanoline, cottonseed oil, lecithin, thiophene, furan, cholesterol, squalene, tygon, latex rubber, and Amberlite XAD-2. Other compounds giving vigorous reactions were pyrrole, sodium tetraphenylborate, tetraphenylarsonium chloride, 1-nitrous-2-naphthol, 2 hydroxyquinoline coumarin, Amberlite CT-120, quinozaline, and anthranilic acid. The authors speculate that the violent reactions encountered with these substances could probably have been avoided if a greater excess of nitric acid had been used.

Table XXVI is a list of the compounds studied by the authors showing the percentage of carbon left after nitric–perchloric acid treatment. This table is a handy guide to workers dealing with samples containing compounds of these types.

SELECTED PROCEDURES FOR SAMPLE DISSOLUTION

This section provides a range of sample decomposition procedures found generally useful in the author's laboratory. Among these, the analyst may wish to choose the best one for his sample type. In subsequent sections, complete procedures containing dissolution and analysis details for specific sample types are also given.

Pressure Method for Wet Digestion of Biological Materials (113)

Comment on the Method

This procedure, because it employs pressure, results in a more complete dissolution compared to conventional methods. It can be used on samples that are to be analyzed for mercury. Fish, bird, and plant tissues have been successfully dissolved. Some samples containing lipids do not yield a clear solution by this method. Despite this, the cations calcium, magnesium, copper, iron, zinc, sodium, potassium, and lead are quantitatively extracted (114).

Being low temperature and a closed system, this decomposition method is well suited to the analysis of volatile metals. Technician time is also kept to a minimum.

Procedure

A 5 g sample is placed in a 2 or 4 oz Nalgene bottle with 1 ml perchloric acid and 2 ml nitric acid, sealed tightly, and allowed to stand to predigest overnight. Samples are then placed in hot running water for 2–3 hr, and then cooled. The caps are removed, 2–3 ml distilled water added, and samples are reheated in hot water in a fume hood to expel the excess acid. The ratio of 1:2 perchloric:nitric acid seems to be the lower limit for good digestion.

Table XXVI

Analysis for Residual Substances after Wet Oxidizing Various Samples with Perchloric and Nitric Acids

Sample	C (%)[a]	Residual substances[b]
Amino acids and proteins		
Glycine	20	Glycine (V), NH_4C1O_4 (X)
Alanine	67	Alanine (V), acetic acid (V), NH_4C1O_4 (X), unknown (W)
Serine	0	NH_4 + (V)
Threonine	0	NH_4 + (V)
Leucine	3	leucine (W), NH_4 + (V)
Phenylalanine	0	NH_4Cl) (X)
Tyrosine	0	NH_4 + (V), trace unknown (W)
Hydrosyproline	2	NH_4ClO_4 (X), unknown (W)
Proline	52	Proline (V,X) NH_4ClO_4 (X), unknown (W)
Cystine	0	NH_4 + (V), SO_4[c]
Methionine	52	NH_4 + (V), methanesulfonic acid (V,Z)
Histidine	9	Histidine (W), NH_4Cl_4 (X)
Tryptophan	0	NH_4ClO_4 (X), trace unknown (W)
Glutamic acid	14	NH_4ClO_4 (X) acetic acid (V), unknown (W)
Lysine	42	NH_4ClO_4 (X), unknowns (V,W)
Arginine	0	NH_4ClO_4 (X)
Albumin	0	NH_4ClO_4 (X)
Gelatin	1	NH_4ClO_4 (X), unknown (W)
Casein	1	NH_4ClO_4 (X), trace unknown (W)
Blood fibrin	1	NH_4ClO_4 (X), unknown (W)
Peptone	2	NH_4ClO_4 (X), unknown (W)
Foodstuff and natural products		
Black pepper	0	NH_4ClO_4 (X), unknown (W)
Cottonseed oil	0	Trace unknown (W)
Cowhide	0	Trace unknown (W)
Dried beef	0	NH_4ClO_4 (X), trace unknown (W)
Dried milk	0	NH_4ClO_4 (X)
Hair	0	NH_4 + (V), trace unknown (W)
Lanolin	0	Trace unknown (W)
Lecithin	1	Unknown (W)
Pine wood	0	Trace unknown (W)
Powdered egg	1	NH_4ClO_4 (X), unknown (W)
Silk	0	NH_4ClO_4 (X)
Soy beans	0	NH_4ClO_4 (X), trace unknown (W)
Tobacco	0	NH_4ClO_4 (X), trace unknown (W)
Wheat germ	1	NH_4ClO_4 (X), unknown (W)
Wool	1	NH_4ClO_4 (X), unknown (W)

Table XXVI (*Continued*)

Sample	C (%)^a	Residual substances^b
Heterocyclic compounds		
2-Amino-6-methylpyridine	0	NH_4ClO_4 (X), trace unknown (W)
3-Aminoquinoline	1	NH_4ClO_4 (X), unknown (W)
2,2'-Bipyridine	59	NH_4 + (V), 2,2'-bipyridine (V,W,X)
2,2'-Biquinoline	42	NH_4ClO_4 (X), 2,2'-bipyridine-5,5', 6,6'-tetracarboxylic acid (V,W)
Furan	0	Trace unknown (W)
Furfural	0	Trace unknown (W)
2-Hydroxyquinoline	0	NH_4ClO_4 (X), trace unknown (W)
8-Hydroxyquinoline	72	NH_4 + (V), quinolinic acid (V,W,Y)
5-Nitro-1,10-phenanthroline	63	NH_4ClO_4 (X), 2,2'-bipyridine-3,3'-dicarboxylic acid (V,W,X)
1-10-Phenanthroline	78	NH_4 + (V), 2,2'-bipyridine-2,3'-dicarboxylic acid (V,W,X)
Pthalazine	0	NH_4ClO_4 (X), trace unknown (W)
Pyridine	88	pyridine (V,W,Z)
Pyrrole	2	NH_4ClO_4 (X), unknown (W)
Quinazoline	0	NH_4ClO_4 (X), trace unknown (W)
Quinoline	72	NH_4ClO_4 (X), quinolinic acid (V,W,Z)
Quinoxaline	0	NH_4ClO_4 (X), trace unknown (W)
Thiophene	0	Trace unknown (W)
2,4,6-Trimethylpyridine	100	2,4,5-Trimethylpyridine (V), chlorodimethyl-pyridinecarboxylic acid (X,Z)
Purines and pyrimidines		
Adenine	3	NH_4ClO_4 (X), unknown (W)
Barbital	14	NH_4ClO_4 (X), acetic acid (V), unknown (W)
Caffeine	26	NH_4ClO_4 (X), methylamine (V,X,Z)
Cystosine	2	NH_4ClO_4 (X), unknown (W)
Guanine	2	NH_4ClO_4 (X), guanine (W)
Uracil	1	NH_4ClO_4 (X), unknown (W)
Uric Acid	0	NH_4ClO_4 (X), trace unknown (W)
Resins and polymers		
Amberlite CG-50	0	
Amberlite CG-120	0	Trace unknown (W)
Amberlite CG-400	19	NH_4 + (V), mixture of methylamines (V,Z)
Amberlite XAD-2	0	Trace unknown (W)
Latex rubber	0	Trace unknown (W)
Nylon	0	NH_4 + (V), trace unknown (W)

(*Continued*)

Table XXVI   *(Continued)*

Sample	C (%)^a	Residual substances^b
Saran	11	Unknown (W)
Tygon	4	Unknown (W)
Selected Organic compounds		
Anthracene	0	
Anthranilic acid	0	NH_4ClO_4 (X), trace unknown (W)
Ascorbic acid	0	
Benzoin oxime	0	NH_4ClO_4 (X), trace unknown (W)
Camphor	4	Camphor (W)
Cholesterol	0	
Cholic acid	3	Unknown (W)
Chrysene	2	Chrysene (W)
Coumarin	0	
Dimethylglyoxime	0	$NH_4 +$ (V)
Nitrilotriethanol	2	$HN_4 +$ (V), unknown (W)
1-Nitroso-2-naphthol	0	$NH_4 +$ (V)
Riboflavin	1	NH_4ClO_4 (X), unknown (W)
Sodium tetraphenylborate	0	
Squalene	0	Trace unknown (W)
Tetraphenylarsonium chloride	0	

^{a} Percentage of carbon content of sample reaming after wet-ashing procedure.
^{b} Identification based upon (V) NMR spectrum of final solution, (W) ultraviolet spectrum of final solution, (X) elemental analysis of isolated product, (Y) infrared spectrum of isolated product, and (Z) mass spectrum of isolated product or fraction.
^{c} Precipitation of $BaSO_4$ on treatment with $BaCl_2$ solution.

Rapid Acid Dissolution of Plant Tissue (115)

Comment on the Method

This method was proposed and tested for the determination of cadmium. In the author's experience it is suitable for most metals in plant tissue. If zinc is to be done, the plastic wrap should be checked for absence of this element.

Reagents and Equipment

Reagent grade chemicals are used. The acid mixture is made from a 2:1 ratio of concentrated nitric:70% perchloric acid.

Deionized water was made by passing distilled water from the laboratory distribution system through a mixed cation–anion exchange resin column.

Procedure

Plant tissue was dried for 48 hr in a forced-draft oven set at 70°C. The plant material was ground in a Wiley Mill using a 40 mesh delivery tube. Some plant tissue required a 20 mesh delivery tube for homogeneity. Samples were stored in an appropriate vial or glass bottle.

Weigh 100 mg of plant materials into 50 ml calibrated test tubes. With an automatic pipet, add 1 ml of acid mixture to the test tubes. Cover the test tubes with a small Pyrex funnel, transferred to the circular digestion rack (Fig. 34) and place on a hot plate. Preheat at 60°C for 15 min or until the reaction subsides, then heat at 120°C until complete dissolution of the sample occurs, which takes about 75 min at 120°C. The entire process is complete in less than 2 hr. Digestion should be conducted in a stainless steel perchloric acid fume hood to minimize the hazard associated with the powerful oxidizing capacity of perchloric acid. After cooling, add deionized water to bring to desired volume. Cover with Saran-wrapped rubber stoppers or appropriate polyethylene stoppers and mix thoroughly. Each sample goes through the entire dissolution process and is brought to volume in the original test tube in which the plant tissue was weighed and digested.

Soluene Method for Tissue Solubilization (116)

Comment on the Method

A quaternary ammonium hydroxide tissue solubilizer is used. This organic-based material has the distinct advantage that it enhances the absorbance signal, for the analyte, compared to that obtained in aqueous digests. The procedure has been tested for zinc, copper, iron, and manganese in unnamed tissue. The method of standard additions was used for the subsequent determinations.

Reagents

Reagent grade chemicals were used. The Soluene (Soluene-100, Packard Scientific, Downers Grove, Illinois, USA) contains a 2% (w/v) solution of ammonium-1-pyrrolidene dithiocarbamate.

Procedure

Weigh the tissue sample into a 50 ml volumetric flask. Add 0.5–1.0 ml of Soluene per 100 mg of tissue. Stopper the flask and leave to stand at room temperature for 24 hr. Heating to 60°C will speed solubilization. All tissues investigated gave a clear, homogenous, and aspiratable

Fig. 34. Circular digestion rack (115).

solution suitable for atomic absorption. A three- to fourfold dilution of the preparation is made with Soluene. The samples were analyzed by the method of additions.

Alcoholic Solution of (TMAH) for Solubilizing a Wide Variety of Tissues (117)

Comment on the Method

Alcoholic solutions of tetramethyl ammonium hydroxide (TMAH) are used to decompose human adrenal, aorta, bladder, blood, bone, brain, cecum, fascia, hair, heart, jejunum, kidney, liver, lung, muscle, nails, nodes, pancreas, prostate, skin, spleen, stomach, teeth, testes, thyroid, and urine for flame or electrothermal atomic absorption analysis of cadmium, copper, lead, manganese, and zinc.

The method consists of solubilizing small quantities of tissues with alcoholic TMAH and using various dilutions of these solutions directly in the furnace or by aspiration into a flame. Unfortunately procedural details are sparse. NBS Bovine Liver Standard was analyzed and good recoveries were obtained for all the metals but copper, which gave low results by 20%. Blanks should be run but are usually found to contain negligible quantities of analyte.

Procedure

Place 1 g of tissue, blood, or urine in a borosilicate liquid scintillation vial equipped with a plastic liner. Add 2 ml of 25% TMAH in alcohol (South Western Analytical Chemicals, Inc., Austin, Texas) and heat the vial with shaking in a 70°C water bath for at least 2 hr. Prolonged digestion time causes metal losses. Dilute the clear (usually amber) solutions to 10 ml with deionized water (1:10 dilution).

Using either an Eppendorf pipet or a Hamilton syringe, 5–20 μl of this solution can be used for analysis with the graphite furnace. The elements cadmium, copper, lead, and manganese are analyzed in this manner. For metals with higher tissue concentrations or greater sensitivities such as zinc, the 1:10 dilution solution can be analyzed by using the flame mode.

Use deuterium background correction for all metal analyses. The diluted digest is stable for 2–3 days. Analyze replicate reagent blanks along with the tissue samples. Biological standards are used to calibrate the instrument.

Dry-Ashing Followed by Mineral Acid Decomposition (118)

Comment on the Method

The following procedure is commonly used for plant materials including bark, wood, leaves, stems, seeds, fruit, roots, and highly organic soils. Dry-ashing is done at 450°C. At this temperature mercury will be lost. Arsenic and the other chalconides will most likely be lost in varying amounts. Other volatile metals such as lead, zinc, and cadmium are normally retained. When appreciable siliceous matter is present, e.g., in leaves and stems of many plants, hydrofluoric acid must be included in the acid digestion mixture.

The main advantages of a dry-ashing procedure is that large sample sizes can be used and perchloric acid need not be employed. As emphasized elsewhere, this acid requires the use of a special fume hood and has been known to cause violent explosions when handled improperly.

Procedure

The sample should be broken into small pieces. Although 30 mesh is best, up to 1 mm lengths are permissible. Place the desired sample weight into an appropriate size Pyrex beaker. Place in a large cool oven. Begin heating at a very slow rate. The temperature should rise from room temperature to 450°C in 6–8 hr. Heat at 450°C for 2 hr or until ashing is complete. Then use the aqua regia method given below. If no hydrofluoric acid is necessary, decomposition can be done in the original beaker. If hydrofluoric acid is required use Teflon dishes.

General Procedure for the Determination of Trace Metals in Biological Tissues Using a Flame (118)

Comment on the Method

This procedure is not applicable to all sample types and all metals at all concentrations. It is designed to be a good procedure for a wide range of metals in many of the simpler plants and animals. The analyst should also be acquainted with the other decomposition methods listed throughout the chapter.

Two basic approaches to decomposition both involving nitric acid are given below. In one case perchloric acid is employed, in the other aqua regia. For samples containing appreciable silica, hydrofluoric acid can be included in either mixture. These decomposition procedures may fail to completely solubilize complex samples such as brain or blood. The pressure method, listed above, should be attempted in these cases.

Methods of dissolution involving perchloric acid are usually best. However, many laboratories are not equipped with safety fume hoods to allow use of this acid.

A worker contemplating use of perchloric acid should consult the supplier on precautions to be observed. The following is a short list of the most important:

(1) Always have nitric acid present in excess with perchloric acid at the beginning of a decomposition. Perchloric acid should never be used without nitric acid.

(2) When very easily oxidizable matter is present, a digestion with nitric acid alone should precede the addition of perchloric acid.

(3) A stainless steel fume hood with facilities for washing down the exhaust vents and complete hood insides should be used. There have been many incidents of workers using perchloric acid for years in ordinary fume hoods without incident. Then suddenly, a serious explosion occurs due to the accumulation of perchlorates in the venting mechanism.

The aqua regia (1:3 nitric acid:hydrochloric acid) method does not give as strongly oxidizing conditions as fuming perchloric acid. However, in many instances, this mixture leaches over 95% of the heavy metals out of an organic matrix.

The sample sizes quoted below may be altered. Using 100 ml beakers or Teflon dishes, up to 0.5 g samples can be handled. Acid volumes should be increased but not proportionately, e.g., a 1 g sample requires 25 ml nitric acid and 10 ml perchloric acid.

Dissolutions can often be done in test tubes. Evaporations are difficult in these tubes and hence, the strongly oxidizing conditions obtained by fuming perchloric acid, do not materialize. However, for easily dissolved samples, test tubes are satisfactory. Heating of test tubes can be done in a sand bath, aluminum block, or boiling-water bath. The latter is least effective.

The following metals can be analyzed after the above decompositions: copper, manganese, iron, cobalt, nickel, chromium, zinc, cadmium, lead, bismuth, vanadium, molybdenum, antimony, and silver.

In the case of silver, chloride can invalidate the results. As a general rule, silver contents of solutions up to 0.5 ppm can be analyzed without problem in the presence of chloride. When silver is to be done at ppm and higher levels, hydrochloric acid must be avoided. Lead, unless in very high concentration, is not precipitated by chloride. Sulfuric acid is avoided in the procedures to prevent formation of lead sulfate.

Agemian and Chau (89) report the loss of chrome as the volatile chromylchloride when perchloric acid is fumed. Hence, the aqua procedure is preferred for this element.

For most metals, the following method is good only down to the microgram/gram level. It is of great importance to employ background correction, particularly when working near the detection limit.

A lean air–acetylene flame must be employed to avoid interferences in the determination of chromium by several base metals. This, unfortunately, also degrades the detection limit. A nitrous oxide–acetylene flame obviates the interference problem due to base metals.

Several of the base metals interfere significantly with nickel. Again, a lean flame minimizes this problem.

A number of elements interfere in a molybdenum analysis. Preparation of samples and standards containing 1000 ppm aluminum overcomes this problem.

Aluminum interferes in the determination of vanadium. It is best to prepare samples and standards to contain 1000 ppm of aluminum. A fuel-rich nitrous oxide–acetylene flame must be used.

The procedure has been tested on a variety of standard reference materials. Percentage recoveries depend on the element and the sample type. In general, the aqua regia method gives from 80 to 100% recoveries compared to better than 95% by the perchloric method.

Reagents and Equipment

Perkin-Elmer 303, 503, and 603 atomic absorption spectrometers are used in the author's laboratory. The experimental parameters, given in Table XXVII relate to this equipment.

Reagent grade concentrated acids are generally adequate for decompositions prior to flame atomic absorption work.

Calibration standards are made by diluting 1000 μg/ml stock solutions prepared from pure metal or metal salts dissolved in a minimum amount of appropriate acid. Stock and calibration standards are made to contain 1% acid. Multielement standards may be used for most elements.

Procedure

Method 1.   Perchloric, nitric and hydrofluoric acid.   Weigh a 0.2 g sample into a 100 ml beaker. Add 10 ml nitric acid. Heat on medium heat of a hot plate (just below boiling) for 30 min. Top up nitric acid to 10 ml and add 4 ml perchloric acid. Using medium heat evaporate to dryness in a perchloric acid fume hood. Add 1 ml of nitric acid and warm. Add 5 ml of water and filter into a 25 ml flask. Cool and dilute to volume. Make appropriate dilutions, as required, keeping the acid content to 1%.

Table XXVII

Instrumental Parameters[a]

Element	Analytical line (Å)	Slit width (Å)	Flame type[b]	Lamp source[c]
Cr	3579	7	A/A(lean)	HCL
Mn	2795	2	A/A	HCL
Fe	2483	2	A/A	HCL
Co	2407	2	A/A	HCL
Ni	2320	2	A/A	HCL
Cu	3247	7	A/A	HCL
Zn	2139	7	A/A	HCL
Cd	2288	7	A/A	HCL
Pb	2833	7	A/A	EDL
Bi	2231	2	A/A	HCL
V	3184	7	N/A(rich)	HCL
Mo	3133	7	N/A(rich)	HCL
Sb	2176	2	A/A	HCL
Ag	3281	7	A/A	HCL

[a] All elements not indicated otherwise are run with lean flames. In the case of chromium this condition gives poorer detection limits but minimizes interferences in air–acetylene flames. Nitrous oxide–acetylene flames are often used for chromium.

[b] A/A is air–acetylene; N/A is nitrous oxide–acetylene.

[c] HCL is hollow-cathode lamp; EDL is electrodeless-discharge lamp.

If the sample contains siliceous material, the following step should be inserted after addition of 1 ml of nitric acid.

Wash the material from the beaker with a minimum of water into a 100 ml Teflon dish. Scrub beaker walls with plastic stirring rod and rinse. Evaporate to dryness. Add 2 ml hydrofluoric acid, 1 ml nitric acid, and evaporate to dryness. Add 1 ml nitric acid and warm. Proceed as above.

Method 2. Nitric, hydrochloric, and hydrofluoric acids. Weigh a 0.2 g sample into a 100 ml beaker. Add 5 ml nitric acid and 15 ml hydrochloric acid. Place a watchglass over the beaker and digest at medium heat for 60 min. Evaporate to dryness. Add 1 ml nitric acid and evaporate to dryness. Add 1 ml nitric acid and warm. Add 5 ml of water and warm and filter into a 25 ml flask. Cool and dilute to volume. Make appropriate dilutions as required, maintaining the acid content at 1%. If the sample contains appreciable siliceous material, do the above dissolution in Teflon vessels. Insert the hydrofluoric acid procedure above in the same place as for the perchloric acid procedure.

Analysis of Solutions

Run samples and standards using the experimental conditions listed in Table XXVII. Standards should be repeated every 10–20 samples.

It is well to analyze a standard reference sample, e.g., NBS Orchard Leaves, with each sample set. The reference sample is run together with samples and calibrating standards.

Detection Limits

In general, detection limits below 0.2 ppm (solution) are not obtainable.

Determination of Cadmium, Chromium, Copper, Iron, Manganese, Lead, and Vanadium in Biological Materials (119)

Comment on the Method

This electrothermal atomization method was tested on the International Biological Standard (IBS) of Bowen. Excellent agreement was obtained. The author has used the procedure for other standards, e.g., NBS Bovine Liver, obtaining good results.

Decomposition of 0.1–0.15 g samples is accomplished through wet-ashing with a sulfuric acid–hydrogen peroxide mixture. A quartz reaction vessel is used to minimize contamination problems. Up to 10 samples can be ashed at once with a proper equipment set up. Ashing time is 3–5 min. This procedure can be highly recommended.

The author has used Teflon decomposition vessels and has been able to ash up to 0.5 g of sample with increased acid and peroxide volumes.

The elements calcium, potassium, sodium, and phosphorus were added to standards of the metals in amounts that would be present in IBS sample solutions. No interference was noted for cadmium, copper, chromium, iron, and manganese.

Serious depressive effects were noted for this combination of elements on the absorbance of lead and vanadium. Correction for this problem may be made by using the method of standard additions within the linear portion of the calibration curve by preparing a suitable matrix standard.

Reagents and Equipment

A Perkin-Elmer 403 spectrometer was used with a Model HGA 72 graphite furnace. Deuterium arc background correction was employed. Furnace operating conditions are given in Table XXVIII.

Table XXVIII

Heating Conditions

Element (10 μl solutions)	Drying temperature/time (°C/sec)	Thermal treatment/time (°C/sec)	Atomization/time (°C/sec)
Cd	100/30	450/30	2100/5
Cr	100/30	1350/10	2700/5
Cu	100/30	900/10	2500/5
Fe	100/30	1000/10	2500/5
Mn	100/30	1000/15	2500/5
Pb	100/30	600/20	2100/5
V	100/30	1700/10	2700/7

The decomposition vessel is shown in Fig. 35 and is constructed from quartz.

Procedure

A dry sample weighing 0.1–0.15 g is covered with 100 μg/l of concentrated sulfuric acid and heated until fumes of sulfur trioxide appear. About 2 ml of 50% hydrogen peroxide is added dropwise until a clear

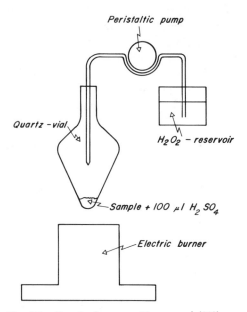

Peristaltic pump

Quartz −vial

H_2O_2 − reservoir

Sample + 100 μl H_2SO_4

Electric burner

Fig. 35. Quartz decomposition vessel (119).

solution is obtained. The solution is cooled and diluted to 10 ml with double distilled water.

The above procedure minimizes the reagents required and the time interval for the digestion. Samples of "dried" biological material stored for a period of time should be redried at 90°C for 20 hr just prior to use.

Standard solutions of the metals are prepared in the range 0.1–10 ng per 10 μl in 0.18 M sulfuric acid. Solutions are stored in quartz flasks.

Run microliter amounts of sample and standard solutions using the conditions given in Table XXVIII. For cadmium, chromium, copper, iron, and manganese use pure solution calibration curves. Because of interference problems use the standard addition approach for lead and vanadium analysis.

ANALYSIS OF BLOOD AND URINE

A good deal of controversy exists about the validity of methods for the very important analysis of lead in blood and other body fluids. Much of this concerns methods of sample preparation particularly when electrothermal atomization is to be used. Some authors specify a surfactant diluent. Others view this step as an unnecessary step likely to increase problems of contamination and high-level nonspecific absorption. Because of this difference of opinion, both reagent-free and surfactant methods are included below. In addition, a flame method and methods involving sampling cups are given. These latter, although they suffer from poor reproducibility, must be included because of their popularity in many hospital laboratories. In general, flame or electrothermal methods are to be preferred over sampling cup approaches.

Lead in Blood by a Reagent-Free Graphite Furnace Method (120)

Comment on the Method

This method was developed in the author's laboratory. It was used as one of the methods in an interlaboratory comparison study involving 30 laboratories around the world. For the standard blood sample, a value of 63.3 compared to a mean of 62.5 μg/100 ml was obtained.

The method is applicable to capillary finger prick samples. Work done in the author's laboratory shows that there is no significant positive or negative bias between venous and finger prick samples from the same patient. The most important point in taking finger prick samples is to thoroughly cleanse the finger by washing, prior to the removal of the blood sample.

Blood samples should not be treated with reagents such as Triton X, as is often recommended as a hemolysis step. This simply adds to already complex matrix problems. Instead the blood samples should all be quick frozen and then thawed prior to the analysis. All analyses should be done as soon as possible after collection. Samples should not be thawed until just prior to the analysis.

Ammonia is added to the graphite furnace *after* the addition of the blood sample. This reagent helps to obtain a more reproducible placement of the blood sample in the bottom of the furnace. Without this precaution, the blood aliquot sometimes sticks on the upper lip of the furnace port. Ammonia also helps minimize frothing of the sample in the furnace.

Background correction can be made using either the deuterium arc or the nonabsorbing line (2820 Å). Problems from background are minimized by diluting blood samples by a factor of 10.

Reagents and Equipment

A Perkin-Elmer 303 atomic absorption spectrometer with the HGA 2000 and a Sargent Welch SRL recorder were used. The 2833 Å lead line was used. Nitrogen purge gas was used at a flow rate of 2.5 liters/min. The gas interrupt was *not* used. Peak height measurements were used as integration does not improve results. Eppendorf microliter pipets were used.

The standard 1000 ppm lead solution was prepared from spec pure lead metal dissolved in the required amount of nitric acid. Dilute working standards were prepared in the range 0–100 μg/liter.

Procedure

Mark the level of blood inside the capillary. Blow sample, as completely as possible, into a dry 10 ml vial with a plastic cap, and dilute with nine parts distilled water, using the original capillary. Swirl solution gently to mix and replace the vial cap. Insert the pipet into a blood sample and depress the plunger all the way. Release the plunger. Inject a 10 μg/liter aliquot by depressing the plunger to the first stop. Replace the plastic pipet tip after each sample. Inject 10 μg/liter of 1:1 ammonia onto the sample inside the tube. Dry the sample, ash, and atomize in the HGA 2000 at 125, 550, and 2300°C, respectively, for 40, 60, and 20 sec. Run standards frequently between samples.

Determination of Lead in Blood Using a Surfactant (121)

Comment on the Method

In this procedure the blood sample is diluted with a surfactant. This was necessary to prevent residue buildup in the graphite tube, which results in a loss of sensitivity. A fivefold dilution was found to be the best compromise between sensitivity and low background absorption.

Interferences from ions present in blood were tested. In lead nitrate solutions a depressive effect on the lead absorbance was noted for sodium and potassium. This was not a problem in blood samples. No interference was experienced with up to twice the concentrations usually used of the anticoagulants heparin, oxalate, or EDTA. The diluent caused no chemical interference. Nonspecific absorption is a problem requiring the use of the deuterium arc background corrector system.

The Perkin-Elmer HGA 2100 used in this work gives higher and more accurate results than the HGA 2000 or 70. This is due to the grooves in the tubes of the former, which prevent spreading of the sample solution along the axis of the furnace.

Accuracy of the method was tested by an intercomparison study with flame atomic absorption. A correlation coefficient of 0.98 was obtained between the two techniques for 102 samples.

Reagents and Equipment

An atomic absorption spectrometer (Perkin-Elmer Model 503) equipped with a deuterium background corrector and a graphite furnace (Perkin-Elmer Model HGA 2100) was used. A wavelength of 2833 Å and a spectral slit width of 7 Å were used. Peak signals were registered on a strip-chart recorder (Perkin-Elmer Model 056). All reference measurements were made with an atomic absorption spectrometer (Perkin-Elmer Model 305A) equipped with a three-slot burner head and an air–acetylene flame.

The HGA 2100 was operated with an internal flow of nitrogen or argon purge gas (normal mode, 15 ml/min, or a setting of 10 divisions on the HGA controller). For determining lead in blood, the following temperature program was experimentally selected as optimum: dry at 100°C for 25 sec, ash at 525°C for 50 sec, atomize at 2300°C for 9 sec.

Deionized water is used throughout. Nitric acid was Ultrex-grade (J. T. Baker Chemical Co., Phillipsburg, New Jersey).

Diluent containing surfactant (Triton X-100, scintillation grade; Eastman Organic Chemicals, Rochester, New York) was used. Prepare a 1 ml/liter aqueous solution.

The lead stock solution (1000 μg/liter) was from Hartman-Leddon Co. (Philadelphia, Pennsylvania).

The lead working standards (0, 50, 100, 150, and 200 μg/liter) were prepared in dilute (5 ml/liter) nitric acid by appropriate dilution of the lead stock solution. With the fivefold sample dilution, the working standards correspond to 0, 250, 500, 750, and 1000 μg of lead/liter in the original sample.

All glassware was acid washed overnight in nitric acid (4 moles/liter) and then thoroughly rinsed with deionized water.

Venous blood samples were collected in heparinized Vacutainer tubes (Becton, Dickinson and Co., Rutherford, New Jersey). Samples were diluted by using Eppendorf microcentrifuge tubes (polypropylene, 1500 μg/liter with snap cap; Arthur H. Thomas Co., Philadelphia, Pennsylvania). Eppendorf microliter pipets were used for both sample solution and injection.

Procedure

Pipet 200 μg/liter of the diluent into a microcentrifuge tube. Add 50 μg/liter of whole blood and mix well. To minimize the transfer error associated with pipetting viscous samples such as whole blood with Eppendorf-type micropipets, flush the pipet tip (blood transfer) several times with the diluted sample.

Inject 15 μg/liter of the diluted sample directly into the HGA 2100 and subject to the above temperature program. Determine the lead content from a standard curve obtained by injecting 15 μg/liter of the lead working standards.

Lead in Blood and Urine: A Flame Method (122)

Comment on the Method

Sample preparation for blood analysis involves precipitation of proteins by trichloroacetic acid followed by an extraction of the lead from the supernatant liquid using ammonium pyrrolidine dithiocarbamate/ methylisobutyl ketone (APDC/MIBK). Simple acidification of the urine prior to APDC/MIBK extraction is the only step required for this sample type.

Kopito et al. (123) did a study of error sources in lead blood determinations. This study showed that this method gives up to 96% lead recovery with one extraction. In this regard it showed distinct advantages over other methods tested. The Kopito study did not involve electrothermal methods.

Reagents and Equipment

Trichloroacetic acid (TCA) (5%) and ammonium pyrrolidine dithiocarbamate (APDC) (1%) were used. Work was done on a Perkin-Elmer Model 214 unit using an air–acetylene flame. The 2170 Å line was employed.

Procedure

Blood.   Add 10 ml of 5% TCA to 5 ml of whole blood. Allow the samples to stand for 1 hr and stir occasionally with a glass rod. Centrifuge the supernatant liquid and decant. Add 10 ml of distilled water to the residue, stir the sample, and centrifuge again. Decant the supernatant liquid and mix with the previous liquid. Adjust the pH to 2.2–2.8 by the addition of about 0.6 ml of 0.5 N sodium hydroxide. Add 1 ml of a 1% solution of APDC and 5 ml of MIBK to this mixture. Shake the samples for 2 min manually or 10 min mechanically, and determine the lead in the organic phase. Use standards containing 0, 0.2, 0.5, 1, and 2 ppm lead in 5% TCA similarly extracted into MIBK. If an emulsion forms when the ketone is added, centrifuge the sample to obtain complete separation of the layers.

Urine.   Acidify 30 ml of urine to a pH of 2.2–2.8 with 5% TCA. Add 1 ml of a 1% solution of APDC and 5 ml of MIBK. Shake the samples for 2 min manually or 10 min mechanically and determine the lead in the organic phase against similarly extracted standards of 0, 0.2, 0.5, 1, and 2 ppm lead. If an emulsion forms at the interface of the layers, centrifuge the sample for 10 min.

Analysis of Lead in Blood Using the Delves Sampling Cup (124)

Comments on the Method

Results obtained by the proposed method were checked against the dithizone colorimetric method. Good agreement was obtained.

When determining blood lead with the Delves sampling cup technique, two (sometimes three) absorption signals are obtained. The combustion products of the partially oxidized sample produce a cloud of smoke that generates an absorption signal(s). The lead peak is preceded by a narrow peak(s) from the combustion products of the sample. The intensity of the smoke peak will vary considerably from sample to sample and also with each determination, due to variations in blood composition and drying characteristics. By using a relatively fast recorder chart speed (60 mm/min minimum) the lead atomic absorption signal can easily be resolved from the nonspecific absorption signal(s). The background corrector simplifies interpretation of the signals. Therefore, its use is desirable though not essential. If a corrector is not available, test for background using the 2803 Å line. The lead signal is read from the base line to the maximum height of the peak. Peak height is proportional to lead content. The signal reaches its peak and returns to the base line in about 4 sec. To separate the smoke peak from the lead peak, the time constant of the electronic system should be less than 0.5 sec.

In the initial experiments it was found that aqueous solutions of lead yielded absorbances about 40% lower than the same lead concentrations in blood. The matrix interference effects probably result from the fact that the crucible operates at a much lower temperature than the air–acetylene flame. Standardization was therefore accomplished by the addition of known concentrations of lead to aliquots of the blood samples. Tracings are obtained by the addition of 10 μl of aqueous solutions containing 0.3, 0.5, 0.8, and 1.0 μg/ml lead to 10 μl of a normal blood sample. The blood samples with additions were dried (140°C) prior to sample oxidation with hydrogen peroxide.

The method of additions is time consuming and troublesome when screening large numbers of samples. An alternative method of standardization is the use of one blood sample, which has been analyzed by the method of additions, as the standard for the direct analysis of other blood samples.

Reagents and Equipment

All data were obtained with Perkin-Elmer Models 290B and 403 atomic absorption spectrometers, equipped with three-slot burner heads and the Delves cup accessory, intensitron lead hollow-cathode lamps and Model 165 recorders. The Model 403 was equipped with a deuterium background corrector. The lead resonance line at 2833 Å and analytical conditions as recommended by the instrument manufacturers were used.

Eppendorf microliter pipets with disposable plastic tips were used for sample handling. The method development was done with a home-made hot plate. However, a Thermolyne Model HPA-1915B is satisfactory.

Procedure

The analytical procedure used in this study, essentially the same as that described by Delves (11) is as follows.

Collect a finger puncture blood sample in a heparinized capillary tube. A finger puncture will easily provide sufficient sample for the blood lead determination. Pipet 10 μl of the blood sample into a nickel crucible using an Eppendorf pipet or equivalent. Dry the sample by placing the crucible on a hot plate at 140°C for about 30 sec. Remove the crucible from the hot plate, cool, and add 20μl of 30% hydrogen peroxide (reagent grade). Return the crucible to the hot plate and heat at 140°C until a dry, yellow residue is obtained. This generally requires about 2 min. During the drying process, a froth will extend above the rim of the crucible but then collapse back into the crucible. At temperatures above 150°C the sample may ignite; therefore, the hot plate used should provide ±5°C temperature regulation. Care must be exercised when handling the crucible to prevent contamination of the sample.

Set the instrument wavelength to 2833 Å. Insert the three-slot burner head and set other instrument controls to manufacturers' recommendations for lead analysis. Install the absorption tube and align to obtain maximum lamp source energy by adjusting the burner controls. Insert a crucible in the wire holder and position it to be 1.5 to 2 mm below the entrance hole of the absorption tube using the rotational, horizontal, and vertical adjustments provided in the microsampling accessory. Condition the crucibles for analytical use by inserting them into the flame several times. To obtain optimum precision, crucibles yielding approximately the same sensitivity should be selected. Set the electronic control to its minimum value, chart speed to 120 mm/min, and recorder full scale to 0.5 A.

Using a suitable micropipet, pipet 10 μl of blood into a crucible and prepare the sample for analysis as described above. Insert the crucible into the holder and introduce into the flame. Remove the crucible from the flame after the absorption signal(s) has been recorded. If the instrument used is not equipped with a deuterium background corrector, the recorder tracing will show several peaks. The peak representing the combustion products of the sample will precede the lead atomic absorption peak. The lead signal is read from the base line to the maximum height on the peak.

At convenient intervals, recorder peaks are measured and sample concentrations computed. The lead content of the samples can be determined by the method of additions or by using one sample that has been analyzed by the methods of additions, as the standard for the direct analysis of other samples. In the latter case, the lead concentration is given by

$$\text{Conc. (sample)} = \frac{\text{peak height (sample)}}{\text{peak height (standard)}} \times \text{conc. (standard)}$$

If noticeable decrease in sensitivity is observed after repeated usage, the crucible should be discarded.

Determination of Platinum and Palladium in Blood and Urine (125)

Comments on the Method

As a precautionary measure, platinum and palladium are being determined in the blood and urine of those assembling automotive catalytic converters. In the following, an electrothermal atomic absorption method is employed.

Certain considerations are necessary for sample preparations, especially in the analysis of blood. Because so little is known about the physiological effects of platinum and palladium and their distribution in the blood, it is considered necessary to analyze whole blood instead of just the serum. This stipulation makes the processing of the blood samples more difficult than if the red cells could be separated from the serum. When the sample is thawed, clotting occurs in spite of anticoagulants, thus rendering the sample inhomogenous.

The samples must be wet-ashed with a nitric-perchloric acid mixture. Ten samples in a batch take 2 hr to process by the decomposition procedure.

No standard samples are available and so the method was evaluated by spiking blood and urine samples with platinum and palladium. The detection limits for platinum and palladium are 0.03 and 0.01 μg/ml in blood and 0.003 μg/ml in urine, respectively.

Reagents and Equipment

A Perkin-Elmer Model 403 atomic absorption spectrometer with a deuterium background corrector, a 10 mV 0.5 sec full-scale recorder, and an HGA 70 graphite tube furnace were used to obtain the data. The 50 μl sample injected into the furnace is dried for 20 sec at 100°C and "charred" for 20 sec at 500°C. With the furnace tilted out of the optical path, the temperature is increased to 1900°C to volatilize the perchloric acid, which otherwise would interfere with platinum and palladium absorbance. The furnace is then tilted back into place, the platinum and palladium are atomized at 2700°C, and their absorbance is recorded as a sharp peak on the recorder.

Standard solutions of platinum and palladium are prepared by dissolving metals of 99.9% purity in aqua regia, evaporating the solution to eliminate most of the nitric acid, and then redissolving in hydrochloric acid.

Procedure

Transfer the well-mixed sample, at room temperature to a 125 ml Erlenmeyer flask. In the case of blood, weigh the container before and after emptying to obtain the sample weight. For urine, weigh the 125 ml Erlenmeyer flask before and after adding the sample.

Add 25 ml of 16 *M* nitric acid and 2 ml of 16 *M* perchloric acid plus one boiling bead to minimize bumping. Heat the solution to fumes of perchloric acid and continue fuming for 5 min with sufficient heat to produce refluxing of fumes halfway up the side of the flask. At the end of this time, approximately 1 ml of perchloric acid remains.

By pipet to the cooled solution, add 5 ml of 1.2 *M* hydrochloric acid and swirl the flask occasionally for 5 min or until all salts are in solution, whichever is earlier. In the case of urine, remove insoluble potassium perchlorate by centrifugation. Transfer the solution to a 2 dram (7.4 ml) vial from which a 50 μl aliquot is taken for injection into the graphite tube furnace.

Prepare calibration solutions by adding aliquots of the standard solutions to a series of 125 ml Erlenmeyer flasks containing either 5 g of platinum- and palladium-free blood or 50 ml of platinum- and palladium-free urine. Carry the solutions through the procedure just as if they were samples. The use of blood or urine, as the case may be, in the calibration solutions is necessary to compensate for the inhibiting effect produced by these substances in the final solution. (The terminology "platinum- and palladium-free" refers to human blood and urine from sources that had no unusual exposure to these metals.) Thus, these samples of blood and urine were found to contain less than the minimum detectable levels of platinum and palladium.

Atomize the calibration solutions before and after each batch of samples. Solutions more than 1 week old are not used.

Blood is collected in evacuated test tubes, e.g., Becton-Dickinson "vacutainer" vials (Catalog No. BD4751), containing sodium heparin. The sample volume is approximately 5 ml. Enough urine is collected to provide a 50 ml sample. Since several days may elapse between collection and analysis, the samples are preserved by freezing.

Determination of Nickel in Blood and Urine (126)

Comment on the Method

The method employs a graphite furnace for the analysis of nickel following its extraction from the blood matrix, using dimethylglyoxime.

Table XXIX

Instrumental Conditions

Nickel lamp current	25 mA
Nickel lamp wavelength	2320 Å
Entrance slit	Position 3
Recorder response	1
Recorder range	10 mV
Recorder setting	0.25 absorbance full scale
Mode of operation	Concentration
Concentration dial setting	5

Problems due to nonspecific absorption negated the direct approach. Nickel dimethylglyoxime is quantitatively extracted into methyl isobutyl ketone (MIBK) in two extractions. Although chloroform is more commonly employed for the extraction, MIBK is used because of the latter's better burning characteristics.

Reagents and Equipment

All measurements were made with a Model 403 atomic absorption spectrometer equipped with a graphite furnace, Model HGA 72 (equivalent to Model HGA 2000 in the U.S.), and Model 56 chart recorder (all from Perkin-Elmer). The instrument was supplied with a deuterium background corrector. Alignment of lamps was carried out as described in the supplier's manual.

Instrument parameters are shown in Tables XXIX and XXX.

After the atomization step, the graphite tube is allowed to cool at room temperature before the next sample is injected.

All reagents were "superpure" grade (E. Merck, Darmstadt, Germany), except as indicated. Diammonium hydrogen citrate was pre-

Table XXX

Graphite Furnace Program

	Temp. control setting	Time (sec)	Digital	°C
Evaporation	1	20	040	120
Decomposition	2	30	100	420
Ashing	3	30	232	1200
Atomization	4	13	850	2600

pared as a 100 μg/liter aqueous solution. Electrolytically pure nickel, 99.95% purity, Falconbridge Nikkelverk was used. Ethanol was purified by redistillation.

Because the sodium salt of dimethylglyoxime gave a high blank, it was purified by recrystallation; 2 g of the salt was dissolved in 500 ml of doubly distilled water. The slightly soluble dimethylglyoxime in acidic solution was precipitated by addition of 25 ml of concentrated hydrochloric acid. The precipitate was separated by filtration and washed with water. The reagent was prepared from this by dissolving 1 g of the purified compound in 100 ml of ethanol.

Whole blood (10 ml) was sampled with use of vacutainer tubes and vacutainer stainless-steel needles (Becton, Dickinson and Co., Rutherford, New Jersey).

Calibration curves were prepared from aqueous, nickel-supplemented serum samples. The stock solution was made by dissolving 0.1 g of electrolytic nickel in 10 ml of nitric acid, followed by dilution to 1 liter, giving a solution of 100 mg/liter. Working standards were prepared by adding 10, 20, and 40 μg of nickel/liter to 3 ml serum samples.

Calibration curves for nickel in urine were prepared by adding 20, 50, and 100 μg of nickel/liter to 3 ml urine samples. After the samples were taken through the complete procedure, the nickel–dimethylglyoxime complex was extracted twice into 3 ml of MIBK.

Procedure

Collect the 10 ml blood sample in a heparinized glass tube, centrifuge, and transfer 3 ml of plasma to a 10 ml quartz crucible. Carefully evaporate the sample on a hot plate and then ash it at 560°C for 5 hr in a muffle furnace. After cooling, dissolve the residue in 3 ml of hydrochloric acid (1 mol/liter), and again evaporate the sample, this time to near dryness, on the hot plate to remove most of the hydrochloric acid. Add 3 ml of water, 2 drops of the ammonium citrate solution, and 200 μl of the dimethylglyoxime reagent, to complex the iron and nickel, respectively.

After adjusting of the pH to 9 with ammonia, transfer the sample from the crucible to a 10 ml centrifuge tube and add 1 ml of MIBK. Shake the mixture for 2 min, centrifuge at 750 g for 2 min to separate the two phases. Aspirate the organic layer with a glass pipet into a 5 ml glass tube and again extract the aqueous layer with 1 ml of MIBK and add this extract to the first one. Inject 50 μl of the mixture into the graphite tube with an Oxford micropipet. Immediately after the sample is injected, start the sequence program on the HGA 72 unit.

Determination of Total and Inorganic Mercury in Urine (127)

Comment on the Method

L-Cysteine is used to complex mercury compounds. Mercury in inorganic compounds, only, is released and reduced at high pH with stannous ion. Total mercury is obtained by using a stannous chloride–cadmium chloride reduction mixture. Organic mercury is obtained by subtraction. The procedure for total mercury was tested against a neutron activation procedure. Good agreement was obtained between these methods.

Storing urine prior to analysis is a problem. The authors find that storing over in a deep freeze results in mercury losses of up to 20% during the first few days. These losses can be prevented by the addition of 1.0 g sulfamic acid and 0.5 ml Triton X-100 detergent per 500 ml of urine or by adjusting the urinary pH to 1.7. In both cases, storage of samples in a cold room (4°C) is convenient.

The detection limit for mercury by the procedures is 0.82 μg/liter.

Reagents and Equipment

The apparatus was assembled as illustrated in Fig. 36. The spectrometer was a Pye-Unicam SP90 Series 1 atomic absorption spectrometer, and a Pye-Unicam mercury hollow-cathode lamp was used as a light source. The burner head was removed and a 15 cm absorption cell positioned in the light path. A Charles Austen Capex Mark II pump was used to generate the flushing air flow and the flow rate was monitored with the air flowmeter on the spectrometer. Absorbance was either

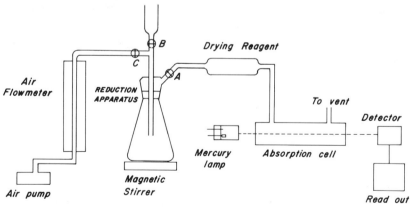

Fig. 36. Apparatus for the determination of mercury in urine (127).

Table XXXI

Instrument Parameters

Mercury lamp current	5–6 mA
Mercury lamp wavelength	2537 Å
Slit width	0.2 mm
Meter response	Position 1
Scale expansion	None
Air flow rate	1–1.25 liters/min

measured on the meter of the spectrometer or was recorded on a strip-chart recorder. Instrument parameters are given in Table XXXI.

The different parts of the apparatus were connected with polyvinyl-chloride tubing, the length of tubing between the chemical reduction vessel and the absorption cell being made as short as possible. The system is "open ended," that is, no circulation of mercury vapor occurs. Magnesium perchlorate, used as a drying agent, was placed in the bulb of a 10 ml glass pipet. Cotton wool was loosely packed in the tubing between the reduction and absorption cells. Both the cotton wool and magnesium perchlorate must be changed at weekly intervals or once every 60–70 analyses.

A special reduction apparatus was constructed for the generation of mercury vapor (Fig. 36). Reaction flasks of 50 ml capacity were used. A 5 ml dropping funnel was used to add the sodium hydroxide solution that initiated the reduction reaction. Stopcock valves A and C were made of polytetrafluoroethylene Pyrex 4 and valve B of polytetrafluoroethylene Pyrex 2. The ground-glass "quick fit" joint size was 14/23. The taps on the reduction cell are closed during the conversion of mercury compounds to atomic mercury to allow maximum partition of the mercury vapor before it is flushed through the absorption cell.

All reagents were of "analar" grade. Deionized water was used in the preparation and dilution of all solutions.

L-Cysteine (or *L-cysteine HCl*).   Dissolve 10 g in 1 liter of 1 mole/liter nitric acid solution.

Stannous chloride solution.   Dissolve 100 g in 1 liter of 2 moles/liter hydrochloric acid solution.

Stannous chloride (100 g/liter), cadmium chloride (100 g/liter) reduction solution (for organic mercury).   Dissolve 100 g stannous chloride and cadmium chlorides in 1 liter of 2 moles/liter hydrochloric acid solution.

Sodium hydroxide.   Prepare 300 and 450 g/liter solutions.

Sulfuric acid.   Prepare 1:9 and 1:1 solutions.

Magnesium perchlorate.   Use anhydrous granules.

Stock inorganic mercury solution.   Dilute "atomic absorption spectroscopy grade" (British Drug House) mercury chloride solution (1 mg Hg/ml) to a concentration of 10 μg/ml with deionized water. This solution is stable for several weeks. Each day, prepare working standard solutions in the range of 10–50 μg Hg/liter by dilution of this stock solution.

Stock organic mercury solution.   Dissolve 17.1 mg of methyl mercury iodide or 16.8 mg of phenyl mercuric acetate (equivalent to 10 mg of mercury) in a small volume of methanol. Transfer the solution to a 100 ml calibrated flask containing 20–30 ml of deionized water, and dilute to the mark with water. Prepare working standard solutions daily by dilution of the stock solution with deionized water.

Procedure: Inorganic Mercury

Into a 50 ml quick fit reaction flask, pipet 2 ml of cysteine solution and 5 ml of the urine sample, inorganic mercury standard, or blank (deionized water), and add 0.5 ml of the stannous chloride solution. Connect the reaction flask to the upper part of the apparatus, which should have valves A and C open and valve B closed. Switch on the magnetic stirrer and air pump and adjust the air flow to 1 to 1.25 liters/min. Pipet 3 ml of the 300 g/liter sodium hydroxide solution into the funnel. Turn off the air pump, close valve C, and open valve B to allow the sodium hydroxide to enter the reaction vessel. When all the sodium hydroxide solution has been added close valve A and then valve B in quick succession.

Allow the reduction reaction to proceed for 2 min, then in rapid succession, open valves A and C and switch on the air flow, which should be at 1–1.25 liters/min. Measure the absorbance on the instrument meter or record the peak absorbance as the mercury is flushed through the absorption cell.

Switch off the magnetic stirrer but continue to flush the system with air until the absorbance reading returns to zero. Remove the 50 ml flask from the apparatus and wash it out with the more dilute sulfuric acid solution, tap water, and finally deionized water. The system is now ready for the next sample. About 12–14 analyses for inorganic mercury can be performed per hour.

Procedure: Total Mercury

Into a 50 ml quick fit flask, pipet 2 ml of the cysteine solution and 5 ml of urine, organic and (or) inorganic mercury standard, or deionized water blank, 1 ml of the 500 ml/liter sulfuric acid, and 1 ml of the stannous chloride–cadmium chloride reduction solution. Proceed as described for inorganic mercury analysis, except that 3 ml of 450 g/liter sodium hydroxide should be added instead of 3 ml of 300 g/liter sodium hydroxide. About 10–12 analyses for total mercury can be performed per hour.

Determination of Aluminum in Whole Blood (128)

Comment on the Method

A direct method and a wet-ashing method for the determination of aluminum in whole blood were investigated. In the direct method, a standard blood sample was used for calibration. Aqueous solutions of aluminum could be used for calibration using the wet-ashing technique.

The direct method has the advantages that no reagent addition is necessary and no separations or preconcentrations need be done. This method was tested in the original author's laboratory against the wet-ashing procedure and results found to be comparable. Two samples were analyzed independently by another laboratory and good agreement obtained.

The detection limit of the method is 0.05 μg/ml with a relative standard deviation of 8% at 0.35 μg/ml aluminum.

Reagents and Equipment

The measurements were made with a Perkin-Elmer 400S atomic absorption spectrometer, the cup and tube version of the Varian-Techtron CRA 63 resistance-heated graphite furnace, and a Perkin-Elmer single-channel recorder. The instrument was equipped with two lamps for background correction; however, as apparent from the preliminary investigations, the analyses can be made without using a background corrector.

The furnace was heated by a four-step power supply constructed in the laboratory. Temperature calibration curves for the furnaces were recorded with an Ircon radiation thermometer.

Sample and standard solutions were introduced into the cup with a 2 μl ultramicropipet (Oxford Laboratories Inc.), and a 5–50 μl adjustable Finnpipette (Kemistien OY, Finland).

For the decomposition of whole blood, 100 μl tubes with stoppers were made from rods of a dense quality of polytetrafluoroethylene (PTFE). During heating the tubes were placed in holes drilled in a circular steel plate, and the stoppers were pressed into the tubes by a second plate fastened to the perforated plate with a screw in the center. The tubes were heated in a drying oven.

A primary 1000 μg/ml aluminum standard solution was prepared by dissolving 1 g of high-purity metal (Johnson Matthey, London) in 200 ml of (1:9) sulfuric acid (Suprapur, Merck), and diluting the solution to 1 liter with deionized water.

Secondary standard solutions were prepared daily by dilution.

Concentrated nitric acid (Suprapur, Merck) was employed for the decomposition of blood.

The graphite furnace was purged with argon of purity 99.9% (by volume).

Procedure

Before the start of the measurements, warm up the hollow-cathode lamp for about 15 min. Adjust the flow of argon to 4 liters/min. The measurements are made at 3093 Å. All analyses are based on measuring peak heights. Transfer undiluted whole blood (2 μl) to the graphite cup of the atomizer; during the introduction of the sample, the pipet tip should not touch the interior walls of the cup. Analyze the samples with the following heating program: drying at 90°C for 40 sec, first ashing at 400°C for 30 sec, second ashing at 1200°C for 30 sec, and atomization at 2500°C for 7 sec.

Two blood samples are employed as standards, the content of aluminum in these being established by the standard addition technique. To three of a series of four 1 ml portions of the blood to be used as standard, 10 μl of 10, 20, and 50 μg/ml aluminum standard solutions are added; the concentrations in the blood of the aluminum added thus corresponds to 0.1, 0.2, and 0.5 ppm, respectively.

Determination of Calcium in Serum (129)

Comment on the Method

The following is a "referee method" evaluated in eight independent laboratories. If the method is used as described, calcium in serum can be determined to within ±2% of the true value over the concentration range 1.5–3 mmole/liter. Because of the length of time required and the volume of sample needed, the method is not suitable for routine use. It is, however, most useful in assessing accuracy of routine methods. All details of the procedure must be adhered to closely.

Reagents and Equipment

Reagent specifications. Water (preferably distilled and deionized) should measure at least 10^6 Ω specific resistance at 25°C. It should be available in large quantity for use as a diluent and for the final rinse operation on all glassware and apparatus coming in contact with the solutions involved. Only water that meets these specifications is to be used in these operations.

Calcium standard solutions should be prepared from a calcium carbonate issued and certified by the National Bureau of Standards. Its identification number is SRM 915. This material should be dried for 2 hr at 140°C and cooled to room temperature in a desiccator before use.

Lanthanum oxide should be of high purity and known to contain less than 15 μg of calcium per gram (15 ppm).

Sodium, potassium, and strontium chlorides should be ACS analytical reagent grade (ACS-A.R.) quality. These materials should be dried before use (see above).

Hydrochloric acid meeting ACS-A.R. specifications should be used.

All glassware (10 ml volumetric pipet to contain 500 ml volumetric flasks, etc.) could meet NBS class A specifications. All glass or plastic surfaces coming into contact with reagents, water, diluent, or sample must have been previously cleaned as follows:

Use routine cleaning procedure (hot water with detergents, plus usual rinses). Soak glassware overnight in HCl (1.0 moles/liter). Rinse with several portions of distilled water (5–6 minimum). Air-dry (inverted) in a dust-free environment.

Stock blank solution. 140 mmole of sodium chloride and 5.0 mmole of potassium chloride per liter. To a clean 1 liter volumetric flask, add 8.18 g of sodium chloride and 373 mg of potassium chloride. Dissolve in water and fill to the neck. When at working temperature (ambient), dilute to the calibrated volume and mix by inverting the flask six times, let stand for several minutes, invert a further six times, and again twice immediately before reading.

Diluent solution (10 mmole of lanthanum chloride and 50 mmole of hydrochloric acid per liter. Plan to make sufficient diluent for the work to be performed in one continuous series.) Transfer 1.63 g of La_2O_3 to a 1 liter flask and dissolve in 10 ml water and 6.7 ml of concentrated hydrochloric acid. After the lanthanum oxide is dissolved, dilute with water to the neck of the flask. When the solution has reached ambient temperature, dilute to calibrated volume and mix by inverting the flask as above. (*Note:* If an internal reference is to be used, add 30.6 mg $SrCl_2 \cdot 6H_2O$/ liter.)

Standard stock solutions of calcium.   Prepare a minimum of three concentrations at 2.00, 2.50, and 3.00 mmole Ca/liter with each to contain 140 mmole of sodium chloride and 5 mmole of potassium chloride per liter. To each of three 1 liter volumetric flasks, add 8.18 g of sodium chloride and 373 mg of potassium chloride. To the first flask (2.00 mmole Ca/liter) add 200.2 mg of $CaCO_3$, to the second flask (2.50 mmole Ca/liter) add 250.2 mg of $CaCO_3$, and to the third (3.00 mmole Ca/liter) add 300.2 mg of $CaCO_3$. To each flask add a few milliliters of water and 1 ml of concentrated hydrochloric acid. Make sure all the calcium carbonate is in solution before diluting with water to the neck. When at ambient temperature, dilute each flask to calibrated volume and mix by inverting the flask 30 times. Label all flasks appropriately.

Procedures

Dilution procedure.   All solutions should be at a relatively constant ambient temperature.

All solutions except the diluent, but including the unknown samples, are diluted fiftyfold, by using a 10 ml volumetric (to deliver) pipet and 500 ml volumetric flasks. Only one 10 ml volumetric pipet is to be used throughout, to reduce errors caused by differences in drainage times between the aqueous or dilute acid and sera solutions, and to eliminate errors caused by volumetric differences between pipets.

Transfer to a 500 ml volumetric flask approximately 450 ml of diluent stock solution. Add to the flask 10.00 ml of the blank stock solution by using the 10 ml pipet. After the pipet stops draining, gently blow out the residual liquid by using a rubber bulb. Rinse the pipet three times with diluent from the flask, each time returning the pipet contents to the flask by drainage and blowing. Dilute to calibrated volume with diluent and mix thoroughly with 30 inversions. Set aside. Into the pipet, aspirate water contained in a clean beaker. Fill to slightly above the mark and discard. To condition the pipet, fill to 1–2 mm above the mark with 2.00 mmole Ca/liter standard stock solution. Discard. Repeat this step twice.

To a 500 ml volumetric flask, transfer 10.00 ml of the 2.00 mmole Ca/liter standard stock solution by the technique described above.

Repeat, but using the 2.50 and 3.00 mmole Ca/liter standard stock solutions. Condition the pipet each time, using the appropriate standard stock solutions.

After the blank and standard solutions have been diluted and the pipet rinsed with water, draw 2–3 ml of the first unknown solution or serum into the pipet. Place a finger over the end of the pipet and then withdraw from the unknown solution container. Tilt the pipet to a horizontal position and slowly rotate the pipet to wet thoroughly all internal surfaces. Allow a small amount of air to leak past the finger so that the rinse solution may come into contact (a small way above the mark) with the upper stem surface. Discard. Repeat the rinse and conditioning operation once more.

Fill the pipet to the mark with the unknown solution (or serum) and deliver into a clean 500 ml volumetric flask. (*Note:* The tip of the pipet should remain in the diluent to prevent the formation of foam by the serum.) Rinse the pipet three times with diluent solution, returning all rinses to the flask. Dilute the flask to calibrated volume with diluent solution and mix by inverting 30 times.

Rinse the pipet with water, condition with the next unknown solution (or serum), and repeat the above as many times as there are unknowns to be analyzed. [*Note:* If the rinse solution does not completely wet the sides of the pipet, clean the pipet by rinsing it several times in a mixture of strong acid (HCl:HNO_3, 3:1 by volume).] Then, repeat the rinse.

At the conclusion of the dilution procedure there should be

(a) One 500 ml volumetric flask containing a fivefold dilution of the blank stock solution. Label B.

(b) Three 500 ml volumetric flasks each containing fiftyfold dilutions of the calcium stock standard solutions. Label 2.00, 2.50, and 3.00 (mmole Ca/liter).

(c) As many 500 ml volumetric flasks, each containing fiftyfold dilutions of each of the unknown solutions (or sera) as there are to be analyzed. Label appropriately. (*Note*: For the sake of clarity in the following steps, it is assumed there is one unknown labeled X.)

Atomic absorption procedure. Acetylene of highest purity should be used. It should be withdrawn slowly so that acetone in the porous filter will not be drawn off. The air should be passed through a water trap and filter. Instability caused by the following must be avoided: 1. Main-line voltage fluctuation. 2. Noisy nebulizer. 3. Electronic instability. 4. Acetylene pressure below 414 kPa (60 lb/in.^{2}. 5. Moisture in the airline, air currents in the room. 6. Solvent residues in the premixing chamber or drainage siphon.

In general, the full accuracy of the method cannot be attained unless the instrument is in optimum operating condition and meets all the specifications set forth by the manufacturer. Repeatability of readings of the same solution within ±1.0% (maximum) is a necessary condition. The use of integration, the strontium internal reference, and damping, where available, will assist in the attainment of high precision.

Instrument and electrical adjustment. Prepare the atomic absorption spectrometer for operation according to instructions provided in the operator's manual. Place the calcium hollow-cathode lamp in the lamp-housing receptacle. Turn the power supply switch on. Select the optimum current for the lamp, and allow ample warm up time for the lamp to become stable. Adjust the monochromator slit and set the wavelength selector to the calcium resonance line at 4227 Å, or to the instrument's maximum energy setting between 4200 and 4250 Å. Setting wavelength at the actual peak response (true emission line of the lamp) improves both the sensitivity of response and the stability of single-beam instruments. Adjust the photomultiplier diode voltage to give optimum current output with minimum dark current.

Flame condition. Open the tank valves on the air and acetylene supplies. Adjust the secondary regulators as recommended by the manufacturer. Check the burner to make sure the premixing chamber and nebulizer are clean and free of any foreign obstructions. Insert an air–acetylene burner head (either a three-slot Boling or suitable single-slot) on the burner. Light the burner and adjust the air and acetylene flow to rates recommended for the instrument. To stabilize the temperature of the burner head, aspirate water into the flame for at least 10 min before proceeding to the next step. (*Note:* A fuel-rich air–acetylene flame gives optimum sensitivity for the measurement of calcium; however, it may be difficult to obtain the precision specified in this method with a fuel-rich flame. Therefore, it is suggested that a stoichiometric or slightly fuel-rich flame be used to obtain the highest precision for calcium in serum.)

Determination of optimum absorption. Determine the stability and repeatability of the instrument as well as the calibration curve as follows:
Adjust the instrument to zero absorbance while nebulizing water. Nebulize the solution of 2.00 mmole Ca/liter, and measure the absorbance. If the absorbance is not 1.000, adjust the scale expansion of the readout system until the absorbance value is greater than 1.000. (*Note:*

A scale expansion of approximately 5 is required for most instruments.) Readjust the instrument to zero absorbance with water. Nebulize the reagent blank, 2.00, 2.50, and 3.00 mmole Ca/liter and record their absorbances. Nebulize water between each of the standard solutions and check the zero value. Repeat the sequence of blank and standards as outlined above until a repeatability of readings for the same solution is within ±1.0%. Subtract the absorbance value for the reagent blank from the average value obtained for the standard solution. Plot on rectilinear graph paper the absorbance, corrected for reagent blank, as ordinate, vs. the concentration of the calcium standards, as abscissa, expressed in millimoles per liter. (*Note:* If concentration values are determined directly from the instrument instead of the absorbance values, follow the procedure analogously, except use the blank to set the zero concentration.) If the calibration curve is not linear, prepare calcium standards of 2.25 and 2.75 mmole/liter, following the procedure given previously.

Absorbance measurement. Measure the absorbance of the unknown solution as follows: Repeat the calibration curve as outlined above.

Nebulize the unknown solution and then nebulize the two standard solutions that are closest to the value of the unknown. Record these absorbance values. Repeat this sequence of standards and a single unknown until 10 valid measurements have been obtained. Repeat on additional unknowns and their associated standard solutions.

Valid measurement. To obtain a valid measurement, follow the sequence of standards and unknowns and record the data as shown in Table XXXII.

In the preceding example, two valid measurements for the unknown in tests 2 and 4 were obtained, because the difference between consecutive standards is less than 1%. However, the value obtained in test 6 is not valid because the difference between 1.351 and 1.320 is greater than 1%. That is to say,

$$1.355 - 1.354 = 0.001 \text{ or } <0.1\%$$

$$1.486 - 1.480 = 0.006 \text{ or } \phantom{<}0.4\%$$

$$1.354 - 1.351 = 0.003 \text{ or } <0.1\%$$

$$1.488 - 1.480 = 0.008 \text{ or } \phantom{<}0.5\%$$

Calculations

Calculate the sample concentration by an interpolation technique:

Table XXXII

Sequence of Standards and Samples

No.	Sample	2.00	2.25	2.50	2.75	3.00
1.	Standards	1.084[a]	1.222	1.355	1.486	1.608
2.	Unknown (1)			1.477		
3.	Standards			1.354	1.480	
4.	Unknown (2)			1.475		
5.	Standards			1.351	1.488	
6.	Unknown (3)			1.420		
				1.320	1.475	

[a] Typical absorbance values.

Mathematical interpolation. Calculate the concentration in millimoles of calcium per liter for each valid measurement by using the formula

$$C = S_1 + \frac{(A_X - A_{S_1})}{(A_{S_2} - A_{S_1})} (S_2 - S_1)$$

where C is the sample concentration (mmole Ca/liter), S_1 the concentration of the lower standard (mmole Ca/liter, S_2 the concentration of the upper standard (mmole Ca/liter, A_X the absorbance of the unknown, A_{S_1} the absorbance of the lower standard, and A_{S_2} the absorbance of the upper standard.

Graphic interpolation. Plot the absorbance values for the upper and lower standards vs. concentration (mmole Ca/liter) on rectilinear graph paper for each set of data. Draw a straight line between data points and determine the concentration of the unknown from the curve.

Computer interpolation. Use a least-squares plot of the upper and lower standards to compute the concentration of the unknown with a computer.

Determination of Submicrogram Amounts of Mercury in Standard Reference Materials (130)

Comments on the Method

Rains and Menis deviate from the traditional cold-vapor absorption tube technique. They recommend heating the absorption tube to 200°C to avoid moisture condensation problems in the cell. The authors note literature references to light-scattering interferences due to the presence of moisture in the mercury vapor stream. To correct for this problem and spurious absorption by other entrained vapors, they utilize a background correction procedure. Interestingly, however, they found that the heating step eliminated any background effect due to water vapor.

Interference from a variety of anions was tested. Little adverse effect was noted except when a reducible ion such as $Cr_2O_7^{2-}$ was present in sufficient quantity to consume appreciable reducing agents. It is important to note in relation to this latter point that the authors eliminated the use of hydroxylomine hydrochloride as a prereduction step prior to mercury reduction with stannous chloride, because this material "was a constant source of mercury contamination resulting in a high reagent blank."

Samples of the standard reference materials (National Bureau of Standards) Liver, Orchard Leaves, and Coal were analyzed by the proposed procedure and by neutron activation, and excellent agreement was obtained.

Reagents and Equipment

Mercury standard solutions. Dissolve 1.000 g ultrahigh-purity mercury (99.99%) in 0.1 N nitric acid and diluted to 1 liter in the same solvent. Prepare a fresh working solution daily by transferring 10.0 ml stock solution to a 1 liter volumetric flask and diluting to volume with 0.1 N nitric acid. Transfer 1 ml of this solution to a 1 liter volumetric flask and dilute to volume with 0.1 N nitric acid.

Reducing solution. Add 300 ml of water and 50 ml sulfuric acid to a 500 ml volumetric flask. Cool the contents to ambient temperature. Add 15 g of sodium chloride and dissolve. After sodium chloride is completely dissolved, add 25 g of stannous chloride. Shake the solution until the stannous chloride is completely dissolved. Dilute the contents to volume.

Diluting solution. Prepare an aqueous acid solution 1 N in nitric acid and 2.4 N in sulfuric acid, using ultrahigh-purity acids. Check the mercury content on each batch. (Only acids that contain <0.5 ng Hg/ml can be used.) Distilled water passed through a mixed ion exchange column resin bed prior to use is employed.

Atomic absorption spectrometer. A double-beam (ratio method) instrument, consisting of 0.5 and 0.25 m Ebert mount monochromators (Jarrel-Ash) with gratings blazed for 3000 Å, beam splitter, and mechanical chopper was used. Electronic circuitry consisted of four units (0–2100 V multiplier phototube power supply, selective amplifier and synchronous detector, ratio converter, and DCR-2 digital readout), and chart recorder.

In the ratio mode the two channels are used, one for signal A (mercury line at 2537 Å) and the other for signal B (aluminum line at 2652 Å). Signal B is used as reference for obtaining A/B ratio. The input of each channel was derived from a multiplier phototube powered by a single DC power source. The output of the photodetector was amplified and fed into the digital readout. In the absorption mode, the outputs of two filter–amplifier stages provided two potentials for ratio measurement.

The mercury hollow-cathode lamp was powered by a stable power supply, operated at 8 mA. The aluminum signal was obtained from an aluminum–mercury amalgam on an aluminum sleeve in the lamp. Any commercially available hollow-cathode lamp capable of providing both mercury and aluminum signals is acceptable.

The absorption cell was constructed from 30 × 0.8 cm i.d. Vycor tubing with gas inlet and outlet ports and quartz windows. Quartz windows are secured to the Vycor tubing with epoxy cement. The cell is wrapped with heating tape to maintain a surface temperature of 200°C.

The reduction cell consisted of a 125 ml gas-washing bottle with coarse fritted bubbler. Purge gas was introduced into the reduction cell through a flowmeter (0.3 liters/min). Teflon tube 1/16 in. i.d. was used to connect the reduction cell to the absorption cell (length of connecting tube was not critical).

The digestion flask was a 250 ml flat-bottom boiling flask with 24/40 joint. A 300 mm Allihn condenser filled to 50 mm with Raschig rings and glass beads was fitted to the flask.

The apparatus is shown in Fig. 37.

Calibration is accomplished as follows. (*Note:* The ratio mode is used for all samples in which an interferent is suspected. If no interferent is present, the single beam system may be employed.)

In the ratio mode, the 0.5 and 0.25 m monochromators are adjusted to read 100% transmission at mercury 2537 Å and aluminum 2652 Å lines, respectively. For the single-beam system, only the 0.5 m monochromator is adjusted to read 100% transmission at the mercury 2536 Å line. Heat the absorption cell to 200°C and adjust the argon flow through reduction cell and absorption cell to 0.25 liters/min. Transfer aliquots of standard mercury working solution containing 0.01–0.2 μg mercury to the reduction cell and dilute to 25 ml with the diluting solution. Add 20 ml reducing solution and close the system immediately. Record maximum absorbance and minimum ratio value of the mercury line

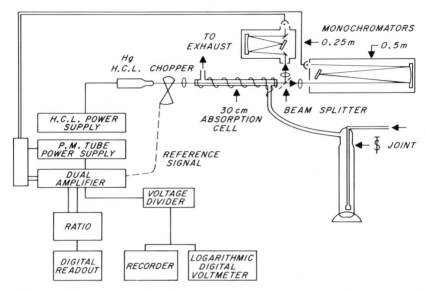

Fig. 37. Equipment for determination of mercury in standard reference materials (130).

from digital readouts. Plot the absorbance and/or ratio values obtained independently against micrograms of mercury to give a linear calibration.

The absorbance signal appears immediately and reaches a maximum within 20–30 sec. It remains constant for about 5 sec and then slowly decreases to zero. The time required for the signal to return to background under these conditions is about 3 min. To speed up the return of signal to baseline, the aeration tube is removed from the sample solution reduction cell after peak absorbance has been reached. The aeration tube is inserted into a clean flask containing dilute nitric acid. Analysis is completed within 1 min if this technique is used.

Procedure

Transfer a 2–3 g sample to a digestion flask. Add 25 ml of 18 N sulfuric acid and 20 ml of 7 N nitric acid. Connect the Allihn condenser and flush with cooling water. Using a heating mantle gently heat the flask for 1 hr at 150°C. Cool the flask and add 20 ml nitric–perchloric acid through the condenser. Drain the water from the condenser and heat the flask until dense fumes of perchloric acid appear. Cool the condenser and rinse with two 5 ml portions of water. Reheat the flask to dense fumes of perchloric acid. Transfer the cooled solution to a volumetric flask and dilute to volume.

(*Note:* For coal samples, digestion time will vary with particle size. The perchloric acid digestion of coal must be done behind a safety shield.)

Determination of Copper, Cadmium, and Zinc in Human Liver Tissue (131)

Comment on the Method

The method is applicable to as little as 5–15 mg of fresh liver tissue and hence may be used for closed-needle biopsy samples. Interference studies were performed and the authors concluded that, although significant effects due to salts in acid solution could be found, these problems were not significant in the actual liver matrix samples as judged by the analysis of NBS bovine liver standard. Hence, matrix standards were not necessary to achieve the accuracy required for liver analysis.

Copper and cadmium were analyzed using the graphite furnace. Zinc was determined using an air–acetylene flame. Because the graphite tube method for zinc is 100 times more sensitive than the flame and zinc is such a ubiquitous element, there can be significant contamination error when the ultrasensitive graphite tube is used to measure this cation. In the original author's laboratory environment, it was impossible to eliminate zinc contamination and so the graphite furnace technique could not routinely be used for this element. In contrast, because of the higher concentrations used, the contamination error with zinc was found to be insignificant when flame atomization was used.

Reagents and Equipment

A Model 303 atomic absorption spectrometer and a graphite tube (HGA 2000) atomizer (both from Perkin-Elmer) were used for all cadmium and copper analyses. Flame atomization with a tantalum three-slot burner (air–acetylene flame) was used for zinc analysis (see Table XXXIII for specific instrumental settings). A 10 mV recorder (Model PWA, Texas Instruments, Houston, Texas) was used for data readout.

All reagents were analytical grade.

For copper and zinc, analytical standards (Dilute-It, J. T. Baker Chemical Co., Phillipsburg, New Jersey) were diluted with 10 mmole/liter nitric acid to a final concentration of 10,000 μg Cu and Zn per liter. Working standards were made by diluting an aliquot of primary standard with 10 mmole/liter nitric acid.

The cadmium standard was Harleco American Public Health Association standard (Scientific Products, McGraw Park, Illinois), the pH of which is 1.5. According to information provided by Harleco this standard is made by dissolving cadmium metal in hydrochloric acid. A working standard was prepared by diluting this stock standard with 10 mmole/liter nitric acid.

Samples were injected into the graphite furnace atomizer with 50 and 10 μl syringes (Precision Sampling Corp., Baton Rouge, Louisiana) having a Teflon-tipped plunger, which eliminates measurable copper and cadmium contamination of the sample from the syringe.

Procedure

Portions of human liver tissue are frozen at −20°C until needed for analysis. The portions of liver frozen should be complete cross sections, about 3 cm thick, surrounded completely by capsule.

Before digestion, thaw the liver and strip the capsule away, revealing a clean, uncontaminated surface. Using an acid-cleaned Teflon spatula, excise a 5–15 mg sample, about 5 cm in from the surface.

Space the sample (human liver or NBS Bovine Liver) in a dried, preweighed, acid-cleaned, 3 ml test tube and record the wet weight of the tissue. After this weighing, place the tubes in a beaker on a hot plate and cover the beaker with an inverted larger beaker. The air temperature inside the larger beaker should be about 80°C. Dry the tissue for 24 hr to constant weight, cool in a desiccator, and weigh. About 70% of the wet weight is lost, leaving about 5.0 mg of dry liver tissue. Add 1 ml of 1.0 mole/liter nitric acid and digest the dried sample on a hot plate at 80°C for another 24 hr. Evaporate the 1.0 mole/liter nitric acid slowly on a hot plate and add 2.0 ml of 10 mmole/liter nitric acid with vigorous mixing.

The liver tissue almost always completely dissolves. Incomplete solubilization did not alter the analytical results. Analysis is delayed for 4–6 hr after reconstitution. Cadmium and copper are analyzed using microliter aliquots of the sample solution; cadmium and copper using the experimental conditions given in Table XXXIII. Zinc is determined in the remaining solution (1–2 ml) by flame atomization.

Trace Elements, Zinc, Iron, Manganese, Copper, Lead, and Cadmium in Fish Tissue (132)

Comment on the Method

A standard addition method is used after extraction of the metals from the sample solution using sodium diethyldithiocarbamate

Table XXXIII

Instrumental Settings for Atomic Absorption Spectrometers

	Cu	Cd	Zn
Perkin-Elmer 303			
Resonance line (Å)	3247	2288	2139
HGA 2000 Atomizer			
Purge gas	N_2	N_2	—
Purge gas flow rate			
(liters/min)	5	5	—
Drying time (sec)	10	10	—
Drying temp (°C)	125	125	—
Charring time (sec)	15	15	—
Charring temp (°C)	450	600	—
Atomization time (sec)	10	10	—
Atomization temp (°C)	2500	1950	—
Flame atomizer			
Gas	—	—	Air–acetylene
Air flow (liters/min)	—	—	34
Acetylene flow (liters/min)	—	—	6
Aspiration rate (ml/min)	—	—	5

(NDDC). Recovery studies were made showing between 90 and 104% obtained. The method was not tested on any standard samples. Levels down to 0.4 μg iron, copper, and manganese, 0.2 μg zinc, 0.1 μg cadmium per gram in a freeze-dried sample could be done.

Apparatus and Reagents

A Perkin-Elmer Model 403 atomic absorption spectrometer equipped with a three-slot burner head and a deuterium background corrector was used at the settings shown in Table XXXIV. The microsampling system and a Model 056 recorder were also used, as were Intensitron hollow-cathode lamps.

Table XXXIV

Instrument Settings

Element	Fe	Zn	Mn	Cu	Pb	Cd
Resonance line (Å)	2483	2138	2794	3247	2833	2288
Slit width (Å)	2	7	2	7	7	7
Lamp current (mA)	30	20	20	20	8	8
Scale expansion	4×	2×	4×	4×	8×	8×
Uptake rate (ml/min)	3.6	2.0	3.2	3.2	4.5	4.5

A 5% aqueous solution of NDDC was prepared and extracted twice with 15 ml portions of methyl isobutyl ketone (MIBK) to remove metal contaminants. Reagent grade MIBK was used throughout without further purification. Analytical-grade Merck reagents were used to prepare standards of each element at the level of 1000 μg/ml. All dilutions were made with deionized water. A commercial household pressure boiler was used with test tubes with S.V.L. Joints, No. 611-52, obtained from Sovirel, France.

Procedure

Extreme caution must be taken to avoid contamination of glassware and reagents. All glassware was washed in Thernads mixture (60% distilled water, 30% concentrated hydrochloric acid, 10% concentrated hydrogen peroxide) followed by rinsing in dilute hydrochloric acid, careful rinsing with deionized water, and drying at 90°C.

Freeze-dried samples were prepared from representative samples of muscle, roe (ovaries), soft roe (testis), and liver of cod (*Gadus morrhua*). The freeze-dried samples were homogenized in a mortar and stored in tightly closed jars at room temperature until analyzed.

Weigh 0.25 g of freeze-dried samples into 10 ml Sovirel test tubes and add a mixture of nitric–perchloric acid (1 + 1) to the tubes. Cap the tubes and allow the acids to predigest the samples overnight. Place four replicates of each sample, plus their four respective blanks, in a pressure boiler and heat at 120°C for 2 hr. The solutions should be clear with a slight yellow color. Cool and remove the caps. Add 1 ml of double-deionized water to the tubes and continue the heating in a waterbath for 2 hr so as to expel the excess of volatile acid. Cool the tubes and transfer the contents to 50 ml volumetric flasks. To three of the replicates of each sample and blank add freshly prepared standard solutions of each element, to give the following additions: 5, 10, and 15 μg iron; 10, 20, and 30 μg zinc; 1.0, 2.0, and 3.0 μg copper, manganese, and lead; and 0.5, 1.0, and 1.5 μg cadmium. Dilute the flasks to volume with deionized water.

Remove 40 ml of the sample solutions for extraction. Adjust the pH to 1.3 with dilute sodium hydroxide in a small beaker, followed by the addition of 10 ml of freshly prepared 5% NDDC. This should result in a final mixture of pH 8.0. No precipitation of iron should occur. Extract each replicate in a 100 ml separatory funnel with one 10 ml portion of MIBK. Shake the mixture vigorously for 1 min and transfer the organic phase to a 25 ml volumetric flask and make up to volume with water-saturated MIBK. This solution is used for the atmoic absorption measurement.

Allow the flame and the deuterium lamp to warm up for 15 min before making the measurements. When the water-saturated MIBK is aspirated, fuel and air flows are adjusted so that a stable, nearly non-luminous flame is obtained on the three-slot burner. At this setting the recorder should give an uninterrupted baseline. Maximum sensitivity is obtained by adjusting the burner nebulizer while aspirating a standard solution containing 10 μg Cu/ml of aqueous solution. Apply direct aspiration techniques for iron, zinc, manganese, and copper, obtaining digital readouts in the ten-average mode. For manganese, the absorption should be measured within one hour after the extraction of the complex into MIBK.

The sampling boat system is used for the measurement of cadmium and lead. The system is operated according to the manufacturer's recommendations. The instrument is operated in the absorbance mode, with a setting of 2.0 A full scale and a chart speed of 120 mm/min.

Pipet quantities of 100 and 50 μl to the middle of the boat for the determination of lead and cadmium, respectively. Evaporate the solvent, keeping the boat at the same distance from the flame, for the same length of time for each replicate of the same element. Push the boat into the flame, and record the signal.

Determination of Molybdenum in Plant Tissue (133)

Comments on the Method

An electrothermal atomizer is used. Maximum temperature is required for proper atomization of this high-melting-point element. It is the experience of the author that the maximum temperature of the HGA 70 is barely sufficient for molybdenum analysis.

The proposed method was tested against an accepted thiocyanate colorimetric method. On two alfalfa samples the values obtained were 9.4 and 10.0 μg/g for a high-molybdenum alfalfa and 0.2 and 0.3 for a low-molybdenum alfalfa using flameless AAS and colorimetric methods, respectively. No interferences were found for the plant samples tested.

Reagents and Equipment

The Perkin-Elmer Model 306 atomic absorption spectrometer and the Perkin-Elmer Model HGA 70 graphite furnace were utilized for the electrothermal procedure. Absorption signals were recorded on a Hitachi–Perkin-Elmer Model 159 recorder at 3133 Å with a special slit width of 7 Å. The HGA 70 was operated at program 7. The samples were dried 1.5 sec/μl of sample, charred for 60 sec, and atomized at 10 volts ($\sim$2600°C) for 15–30 sec. Argon was used as the purge gas.

Scale expansions of 10× or 3× are needed for the molybdenum determination.

"Tailing" of the peak is a problem in the molybdenum determination but is overcome by long atomization periods (15–30 sec) at maximum temperature.

Standards of 0.01, 0.02, 0.05, and 0.10 μg Mo/ml are prepared in 5% hydrochloric acid (v/v) and glass-distilled water. A 1000 μg Mo/ml stock solution is prepared by dissolution of 1.840 g of ammonium paramolybdate, $(NH_4)_6Mo_7O_{24} \cdot 4H_2O$, in 1 liter of 10% HCl (v/v) and glass-distilled water. All volumetric ware and storage bottles had been acid washed.

A 20 μl Eppendorf micropipet is used to place all the samples into the furnace.

Procedure

Air-dry plant samples and grind to pass through 20 mesh screen in stainless steel grinding equipment. Weight 1 or 2 g samples into 30 ml, 96% silica crucibles and ash at 500°C for 5 hr. After cooling, dissolve the ash in a 20 ml aliquot of 5% hydrochloric acid. Filter this solution through Whatman No. 50 filter paper and pipet or dilute and then pipet into the graphite cell for molybdenum determination. This matrix is destroyed in about 30 sec at 1100°C on program 7 of the HGA 70. Atomize at maximum temperature for 15–30 sec.

Determination of Chromium in Plants and Other Biological Materials (134)

Comment on the Method

A very important comment is made on the validity of dry ashing. These researchers found that dry ashing left a residue from which chromium could not be quantitatively extracted without use of hydrofluoric acid. This has also been the experience of this author with other elements, e.g., lead.

The procedure thus involves a wet-ashing step with nitric, perchloric, and sulfuric acids. This is used in preference to a dry ash because of the author's worries about chromium losses during the latter process. To prevent possible volatility loss of chromium as chromyl chloride, silver nitrate is added to the wet digestion. Quantitative chromium recovery was tested using radioactive ^{51}Cr. The procedure was tested on the standard samples, NBS Orchard Leaves, and Bowens Kale. Good agreement with accepted values was obtained.

The method utilizes extraction of chromium with 2,4-pentanedione (HAA) into chloroform. The chloroform is subsequently removed by evaporation. Iron, aluminum, and nickel, found to cause interference in the chromium determination by other workers, did not cause any problem in this system.

Samples analyzed include wheat grain, wheat stem, wheat leaf, oak leaf and stem, human serum, and plasma.

Reagents and Equipment

A Perkin-Elmer Model 303, or equivalent, with recorder with at least tenfold signal amplification capabilities, is used. An air–acetylene flame is used with Boling (three-slot) burner 8 mm below center of light beam. Use nitric acid redistilled in glass. Distill about 90% of total 2,4-pentanedione (No. 1088, Eastman Kodak Company, Eastman Organic Chemicals, Rochester, New York), washed or equivalent. Distill this in glass-collecting fraction, boiling from 135 to 137°C, at 745 mm Hg. MIBK (Eastman No. 416, or equivalent) was used. Use methyl orange (0.02% in water). Mix 60 ml acetic acid, 60 ml water, and 60 ml ammonium hydroxide; cool, adjust to pH 5.8, and dilute to 250 ml to prepare buffer. Precipitate potassium perchlorate from perchloric acid with potassium chloride and wash four times with water. Suspend in water and adjust pH to 3–7. Add 10 ml of 2,4-pentanedione and heat overnight at 85°C. Cool, extract twice with chloroform, and filter, washing with water. Ammonium pyrrolidinodithiocarbamate (No. 2081, K & K Laboratories, Inc., Plainview, New York) or equivalent. Prepare washed 1% solution and shake before using. Also prepare chromium solution: Dissolve 0.3736 g K_2CrO_4 in water and dilute to 1 liter. Dilute this solution to a working concentration of 0.1 μg Cr/ml.

Clean digestion flasks by heating beyond fuming with sulfuric acid and potassium perchlorate for 15 min the first time they are used. Fill the flasks in which sample is to be heated overnight with water and 3 ml 2,4-pentanedione and heat overnight. Then wash all glassware with detergent, and rinse in 1 N hydrochloric acid and distilled water before use.

Procedure

Samples may be dried at 100°C without loss of chromium. Samples should not be ground unless essential to subsampling procedures. If grinding is necessary, a Wiley mill equipped with a 20 mesh screen is recommended.

Prepare chromium standard solutions to bracket chromium concentrations in samples and carry through method. (Prepare standard curve even though response to graded levels of chromium is linear over the range 0.0025–0.75 µg Cr/ml MIBK. Samples may be diluted with MIBK to adjust them within range of standard curve provided correction in blank is taken into account.)

Weigh <5 g (dry weight) of material into a 100 ml Kjeldahl flask. Add two glass beads, 2 ml sulfuric acid, and 10 ml nitric acid (20 ml for samples >1.5 g). Heat until brown fumes begin to clear. Cool slightly and add 0.75 g potassium perchlorate and evaporate to fumes. If digest darkens before appearance of perchloric acid fumes, immediately remove flask from digester unit, add 1 ml nitric acid (1 + 1), and continue heating. Repeat if necessary. Let digest cool and add 5 ml water. Transfer to separatory funnel (Teflon stopcock) and add 0.5 ml of 10% sodium sulfide. Let contents stand approximately 10 min so all hexavalent will be reduced to trivalent chromium. Add 2 drops of methyl orange and ammonia to yellow color and then hydrochloric acid to red color (pH~3). (This should be done quickly, since indicator has tendency to fade.) Cool to ~4° before adding 3 ml 1% ammonium pyrrolidinodithiocarbamate and 2 ml 2,4-pentanedione. Shake (room temperature) for 90 sec and extract with two 5 ml portions of chloroform. Shake 1 min per extraction. (This stripping step is the only operation that must be carried to completion once it has been initiated.) At low temperature trivalent chromium is kinetically slow to react with many complexing reagents. However, this step should be carried out as rapidly as possible. Transfer aqueous phase to reagent bottle, add 1 ml acetate buffer and 3 ml 2,4-pentanedione, cover bottle mouth with watchglass to prevent evaporation and contamination, and heat 16 hr in constant temperature oven at 90°C. Cool to <15°C and extract twice for 5 min with 5 ml chloroform. Collect organic phase in 45 ml evaporating dish and evaporate.

Dissolve chromium in MIBK by washing soluble contents into calibrated tube. Dilute to final volume of ≥4 ml and mix. Aspirate into atomic absorption instrument, reading at 3579 Å line. Adjust flame to be reducing with flow rate of 5–6 ml MIBK/min.

ANALYSIS OF FOODS

Determination of Traces of Lead and Cadmium in Foods Using the Sampling Boat (135)

Comment on the Method

This method employs a redesigned sample boat mount to eliminate breakage problems with boat arms, which occur after several insertions of the boat into the flame.

A temperature of 500°C is used for ashing. The author's experience suggests that this temperature is very near, if not above, the temperature at which volatility losses of cadmium occur from some sample types. The worker is hence cautioned to either use a slightly lower temperature, e.g., 450°C, or test for losses from the sample type being analyzed.

No interference studies have been done in connection with the proposed procedure. However, the results obtained by the method are compared to those obtained by conventional flame atomic absorption and acceptable agreement is recorded. Also additions of lead and cadmium were made to the food and the recoveries of the additions were satisfactory.

Reagents and Equipment

A Perkin-Elmer Model 403 atomic absorption spectrometer equipped with a microsampling accessory, a three-slot burner, a deuterium background corrector, and cadmium hollow-cathode lamps, 1 mV strip chart recorder, and air–acetylene flame was used. A muffle furnace that can be set for 500°C is necessary for ashing the samples.

The boat mount was modified as follows: The rod holding the Delves cup was replaced with two lengths of platinum wire, approximately 4 in. long and 1/8 in. in diameter (other heat-resistant wires may also be used). The wires were bent at one end into a U configuration and at the other end were fastened to the slide by means of screws. The arms of the boat were bent to be parallel with the boat body and folded at midpoints. The platinum wires were bent so that the boat could be suspended on the wires by slipping its arms into the "U" bends of the wires.

Standard solutions of 1000 μg/liter lead and cadmium were prepared by dissolving 1.598 g of lead nitrate, $Pb(NO_3)_2$, reagent grade, in 1 liter of 1% (v/v) nitric acid and 1.000 g of cadmium metal, reagent grade, in a minimum volume of (1 + 1) hydrochloric acid and diluting to 1 liter with 1% (v/v) hydrochloric acid. Further dilutions were made with distilled water just prior to use.

Procedure

Operate the atomic absorption spectrometer according to the manufacturer's recommendations using the absorbance mode and a scale expansion setting of 0.25 A full scale. Set the recorder response control 1 (fastest) and use a chart speed of 240 mm/min.

Mount the boat in the redesigned boat holder and position 6.3 mm above the burner head. While the instrument is operating, push the boat in and out of the flame until no absorption response due to lead or cadmium is observed (new boats produced intense absorption signals). Subsequently, the boats are handled with forceps.

Blend samples of strained baby carrots, mushroom soup, and whole kernel corn in a Waring blender at maximum speed for 0.5 hr. Stir condensed milk with a clean glass rod. Dilute a 10 g portion from each sample to 50 ml with distilled water and mix well.

Add a 0.2 ml portion of the diluted sample to the boat by means of a plastic disposable pipet. Place the boat with sample on an asbestos pad in front of the open muffle furnace, which has been preheated and set at 500°C. Dry the sample cautiously and char (1–2 min) avoiding splattering or foaming over. Place in the muffle furnace and close the door. When white ash is obtained (5–10 min) remove the boat with the ashed sample and place into the boat holder of the atomic absorption spectrometer. While the instrument is operating with the appropriate lamp in place, and the recorder on, push the boat containing the ashed sample into the flame to make the measurement and then pull out of the flame when the recorder pen returns to the baseline. With a plastic disposable pipet, add a 0.1–0.3 ml volume of the standard solution (0.1 μg Pb/ml or 0.01 μg Cd/ml both diluted with distilled water) to the boat and dry by pushing the boat on the slide close to the flame. The response due to the standards is then determined in the same way as for the sample. The concentration of the metal in these samples is calculated by comparing the peak heights or peak areas of the sample with the standards responses.

Detection Limit   Lead, 0.025 μg/g; cadmium, 0.0025 μg/g

Heavy Metals in Meats (136)

Comment on the Method

Cadmium, cobalt, chromium, copper, iron, manganese, nickel, lead, and zinc are analyzed in beef liver and muscle and in turkey muscle. National Bureau of Standards SRM 1577 Bovine Liver was used to assess the accuracy of the method. Good agreement with accepted values was obtained.

Chromium, zinc, and iron are analyzed by flame atomization and the remaining elements using electrothermal atomization. The importance of using the deuterium arc for background correction in the case of the latter is emphasized.

Reagents and Equipment

The samples were analyzed using a Perkin-Elmer Model 360 atomic absorption spectrometer equipped for flame atomic absorption analysis. For electrothermal analyses, a Perkin-Elmer Model 503 atomic absorption spectrometer equipped with an HGA 2100 graphite furnace and a Model 056 recorder was used. Both instruments were equipped with simultaneous deuterium background correctors. Background correction is essential when complex matrices such as these are analyzed with the graphite furnace, but is not always necessary for flame work.

Standards were prepared by dilution of 1000 μg/ml standard solutions purchased from several different suppliers. To obtain accurate results by flame work, standards were prepared with about the same amount of acid as was in the samples. Zinc, iron, and chromium were high enough levels to determine by flame atomic absorption. Several samples contained manganese and copper at high enough levels to determine by flame atomic absorption. The flame methods are significantly faster than the furnace procedures, so whenever the metal levels are high enough to use the flame, it is desirable to do so. Standard instrument conditions as recommended by the manufacturers were used for wavelength, slit width, lamp current, and flame conditions except that a nitrous oxide–acetylene flame was used for chromium.

Initial experiments indicated that severe contamination problems existed. Thus, it was necessary to wash all glassware first with nitric acid. The filter paper was also acid-washed before use.

Procedure

Weigh duplicate samples of about 1 g on a laboratory balance and place in acid-washed beakers. This weight may need to be adjusted depending on metal levels encountered. Dissolve the samples in 10 ml of Ultrex-grade nitric acid (1:1); use the same acid to prepare the standards. Digest the samples on a hot plate in covered beakers at a temperature of about 80°C until only fat remains and the solution volume had been reduced to about 3 ml. Cool the samples, filter, dilute to 25 ml with deionized water, and analyze.

For accurate results using the graphite furnace, the method of additions is used for each sample. To eliminate errors in pipetting, the same pipet is used for samples and standards. Thus, several standards of each element are prepared, and the additions made directly in the furnace. On several occasions, it was noted that not all of the sample was ejected from the tip of the Eppendorf pipets. Therefore, all samples should be pipetted by taking xylene up into the pipet tip first. The plunger is depressed to the second stop, and the pipet tip placed in a container of xylene. The tip is then placed in the sample and the plunger released. Sample is taken up into the tip. The tip is then inserted into the sample introduction port of the graphite furnace, and the sample is expelled into the graphite tube. The xylene sweeps all of the sample into the tube. A new tip is used for each sample.

Table XXXV shows the analytical conditions that were chosen for each element determined in the graphite furnace. In each case, simultaneous background correction is used. Internal gas is kept flowing during the atomization cycle.

For electrothermal analyses, standard instrument conditions are used except for copper, for which the less-sensitive wavelength at 3274 Å is used. For cobalt and nickel, scale expansion of five times may be required in order to work at the low levels of these elements in the samples.

Determination of Trace Metals in Fish Meal and Liver (137)

Comment on the Method

Fish samples are dry ashed at 450°C prior to analysis. The ash is passed through a 270 mesh sieve. Liver and fish samples used for the determination of cadmium were not ashed. The liver sample was ground and passed through 200 mesh sieves. Solid samples are introduced into the furnace. The recommended furnace operating conditions are quite different than those recommended by other authors for these elements. The analyst may wish to experiment with more standard conditions, e.g., Table XXVIII. The accuracy of the procedure is

Table XXXV

HGA 2100 Operating Conditions for Analysis of Meat Samples

	Wavelength (Å)	Dry (°C-sec)	Char (°C-sec)	Atomize (°C-sec)
Cd	2288	150-20	200-15	2100-8
Co	2407	150-30	800-30	2700-8
Cu	3274	150-30	800-30	2700-10
Pb	2833	150-40	600-20	2300-7
Mn	2795	150-30	600-30	2500-10
Ni	2320	150-30	600-30	2700-10

evaluated using NBS (SRM-1577) Bovine Liver, and satisfactory agreement was obtained. Fish samples were also analyzed by a wet decomposition method and results compared with solid sample analysis. Satisfactory agreement was obtained.

Reagents and Equipment

Perkin-Elmer 303 and 400S atomic absorption spectrometers were used; both instruments were equipped with arc source deuterium lamps for background correction, and the 400S spectrometer also had a tungsten lamp for corrections in the 3000–8000 Å range.

Solid samples were atomized in a graphite furnace. These analyses were done with the 303 instrument and were based on measuring peak areas. Liquid samples were atomized in the flame with the 400S spectrometer and the proper background corrector.

Samples were ground in agate mortars and pestles. Weighings were made with semimicro- or microbalances.

The fish meal samples tested were industrial products of mackerel, blue whiting, capelin, and hexane-extracted capelin. The samples had been dried by steam, and no preserving agents had been added. The main constituents of the former three products were protein (~72%), water (~7%), ash (~12%), sodium chloride (~1.7%), and fat (Soxhlet) (~9%). In the hexane-extracted capelin, the protein content was about 80%, and the fat content (Soxhlet) about 0.7%. The samples were relatively coarse powders in which tissue fibers and particles of bone could be seen with a magnifying glass. The samples of fish meal were not dried before analysis.

The Bovine Liver was stored in a refrigerator at about 4°C; before analysis, portions were dried by lyophilization as prescribed by NBS.

Standard solutions of the four elements were prepared by dissolving the appropriate amounts of high-purity metals in a small excess of nitric acid. In all diluted standard solutions the pH was maintained at or below 2.0 by adding, where required, nitric acid. The acids were of Suprapur quality (Merck). The furnace was purged with argon (purity 99.9%, by volume).

The hydrogen peroxide solution (30%) was reagent grade.

Procedure

Transfer about 5 g of the fish meal samples to platinum dishes and char at 450°C in an electric furnace. Mix the residues intimately by grinding in an agate mortar and pestle, to pass a 270 mesh sieve. This ashing step concentrates the trace elements by about 10 times.

Because of the risk of losing cadmium during dry ashing, portions for analysis by direct atomization were taken from the original materials, which were ground to pass a 200 mesh sieve.

The Bovine Liver is not ashed; portions are ground to pass a 270 mesh sieve.

Before the start of the measurements, the hollow-cathode and background corrector lamps should be heated for about 15 min. Adjust the flow of argon through the graphite furnace to 0.36 liters/min. The wavelengths for the measurements are listed in Table XXXVI. Use the following operating conditions for the graphite furnace: (a) dry for 30 sec at 100°C in all cases; (b) ash for 60 sec at 350°C for cadmium and at 450°C for copper, lead, manganese; (c) atomize for 30 sec at 1700°C for cadmium and at 1950°C for copper, lead, manganese; (d) clean at 1950°C in all cases.

Weigh 0.3–6 mg of the solid original or preashed samples in small tantalum scoops and place in the middle of the furnace by means of a specially constructed adjustable inserting device. Reweigh the scoops and move the furnace to its preadjusted position. Atomize two portions of the sample without addition of standard solution; then add three or four portions with 5–20 μl of metal standard solution. From the integrated absorption data, the position of the standard addition curve, the intercept on the abscissa and the standard deviation of the value found can be derived by the method of least squares. Two standard addition-curves are plotted for each metal.

The analyses of the sample solutions are also based on the use of standard addition. From the 50 ml sample solutions, prepared by wet ashing 5 g samples with nitric acid and hydrogen peroxide, transfer four 4 ml portions (for the determination of copper in Bovine Liver, 1 ml portions were taken) to sample bottles. Add to one of the bottles 1 ml of water; to the others add 1 ml of metal solution of varying appropriate concentrations (in the analysis for copper in bovine liver 4 ml of water and 4 ml of 30, 60, and 90 μg/ml copper standard solution were introduced). The series of solutions and blanks are atomized in the air–acetylene flame, and the averages and the relative standard deviations are calculated from the resulting standard addition curve.

Table XXXVI

Wavelengths for Measurement

Metal	Cd	Cu	Pb	Mn
Wavelength (Å)	2280	3274	2833	2795

Determination of Lead, Cadmium, and Zinc in Sugar (138)

Comment on the Method

A novel technique is used to eliminate matrix problems, which would normally be present in the graphite furnace analysis of sugars. The sugar is fermented with yeast to give ethanol and carbon dioxide.

The procedure states that the metals are determined both by direct comparison and by standard additions. The authors conclude, however, that the technique is relatively interference free; a slight interference occurring in the case of lead is overcome by the standard addition technique. Direct comparison may, therefore, be employed for zinc and cadmium.

Recovery studies were used to assess accuracy. The ions were added prior to fermentation. Average recovery was $97 \pm 7\%$.

Reagents and Equipment

The equipment consisted of a Perkin-Elmer Model 306 atomic absorption spectrometer, equipped with the HGA 2000 graphite furnace, a Model 056 recorder, and a deuterium background corrector.

Standard solutions were prepared fresh daily from 1000 ppm commercial stock solutions. The sugars were obtained from the refinery, along with process liquors from various stages of the refining process. The liquors were all about 60% solids by weight.

Procedure

Weigh approximately 5 g of sugar (10 g of sugar liquor) into a 150 ml beaker. Dissolve the sugar in about 35 ml of deionized water, and adjust the pH to 4.5–5 with 10% acetic acid. Add about 0.25 g of baker's yeast to each solution. Place the mixtures in a 40°C oven overnight to complete fermentation. Transfer the solutions quantitatively to 50 ml volumetric flasks and bring to volume with deionized water. Remove yeast by centrifugation and decant the clear supernatant. Prepare yeast blanks containing no sugar.

The operating conditions used to determine each element are given in Table XXXVII. The deuterium background corrector must be used.

A spectral bandwidth of 7 Å and nitrogen purge gas are used. Except in the cadmium determination, nitrogen flow is not interrupted during the atomization stage.

The concentration of metals in the sugar samples is determined both by the method of additions and from a calibration curve prepared using

Table XXXVII

Furnace Operating Conditions

Operating conditions	Element		
	Pb	Cd	Zn
Dry (sec-°C)	30-100	30-100	30-100
Char (sec-°C)	60-400	60-400	60-400
Atomize (sec-°C)	10-2000	8-1600	8-2000
Wavelength (Å)	2177	2288	2140
Sensitivity (pg/0.0044 abs)	8	0.9	0.9

aqueous standards. At least three determinations are made on each sample.

Determination of Calcium, Copper, Iron, Potassium, Magnesium, Manganese, Sodium, and Zinc in Orange Juice (139)

Comments on the Method

Orange juice samples are treated with 10 M nitric acid at room temperature overnight and then at 80°C for 5 hr. Remaining solids are filtered prior to atomic absorption flame analysis. Precision was found to be best for magnesium and worst for manganese at the concentrations found in orange juice. The procedure was not tested on standard samples. No other tests for accuracy were employed.

The author recommends background correction for copper, iron, manganese, and zinc.

Reagents and Equipment

Instrumentation Laboratories atomic absorption equipment was used, the instrumental parameters being those recommended by the manufacturer.

Intensity readings were taken from the digital readout system supplied with the instrument. Three multielement lamps were used: a calcium–magnesium–aluminum, an iron–copper–cobalt–chromium–manganese–nickel–zinc, and a sodium–potassium. All three lamps were purchased from Varian Techtron, Palo Alto, California.

Analytical-grade reagents were used in preparing all standard solutions. Magnesium, iron, copper, and zinc as metals were dissolved in acid and diluted to volume. Sodium, potassium, and calcium standard solutions were prepared from the dry carbonates. The manganese standard was prepared from potassium permanganate. The manganese was reduced to the divalent state with oxalic acid before diluting to final volume. A mixed standard was made from the above reagents to correspond to the following concentrations: 10,000 μg/ml of potassium, 1000 μg/ml each of calcium and magnesium, 10 μg/ml of sodium; 4 μg/ml each of copper, manganese, and zinc, and 10 μg/ml of iron.

Standards simulating orange juice were prepared by making a series of dilutions of 1, 2.5, 5, and 7.5 ml of the above mixture of elements plus dextrose and phosphoric acid in amounts approximately equivalent to those expected in a 10 g sample of frozen concentrated orange juice. In most instances, these dilutions corresponded to the useful working range for the elements under consideration. They were processed through the hydrolysis procedure exactly like the orange juice samples.

Procedure

Weigh individual 10 g samples of frozen concentrated orange juice into 100 ml volumetric flasks. Add 10 ml portions of 10 M nitric acid, one portion to each flask. Agitate the mixtures of sample and acid slowly by rotating the flasks by hand for a few seconds. Allow the flasks to stand overnight in a bath of room temperature water. The cooling effect of the water bath is necessary to prevent excessive foaming during the initial stages of the hydrolysis procedure. The following morning, heat the bath with flasks to 80°C for 5 hr. At the end of the heating period, cool the flasks and contents to room temperature and dilute to volume (100 ml) with deionized water. Remove the solids remaining by filtering the flask's contents through rapid filtration paper into clean, dry plastic bottles. Make dilutions of the filtered matrices 1:1 with water for the determination of copper, iron, manganese, and zinc. For the measurements of calcium, magnesium, sodium, and potassium, use dilutions of 40:1. Make dilutions with deionized water. Acid-wash all glassware and containers prior to use.

Determination of Copper, Manganese, Nickel, Cadmium, Lead, and Zinc in Foods (140)

Comment on the Method

The following foods were analyzed by the proposed procedure: tomato catsup, fresh spinach, canned Bonita fish in oil, apples, shredded wheat, homogenized milk, canned turnip greens, and frozen cod fillets. Accuracy and precision for copper, manganese, cadmium, lead, and zinc were tested by the analysis of NBS Orchard Leaves (SRM No. 1571) and Bovine Liver (SRM No. 1577). Agreement with accepted values was good, average recovery being 96.2%. Precision measured as average standard deviation was 1.4%.

The procedure involves a digestion using nitric acid–sulfuric acid–hydrogen peroxide catalyzed by vanadium pentoxide. Lead is copre-cipitated with strontium and separated by filtration. The other metals are separated from matrix constituents by use of a chelating resin.

Reagents and Equipment

Standards. Stock, 10,000 μg/ml, from the pure metals (Ventron Corp., Alfa Products, Beverley, Massachusetts) were prepared by dis-solving 1.000 g of the metal in hydrochloric, nitric, or sulfuric acids; intermediate, 1000 μg/ml stock standard diluted with 1.6 N sulfuric acid for all metals except lead, where 2 N nitric acid is used (stability approximately 3 months); working solutions, 0–10 μg/ml. The same acids specified for intermediate standards were used; prepared by dilu-tion of intermediate standards at time of use and discarded after use).

Dowex A-1 resin (Chelex 100, 100–200 mesh, sodium form) was from Bio-Rad Laboratories. A pH 6.5 buffer was prepared by adding 140 ml of 0.100 M sodium hydroxide to 500 ml of 0.100 M KH_2PO_4 and diluting to 1:1. The strontium solution (2% w/v) consisted of 6.00 g of $SrCl_2 \cdot 6H_2O/100$ ml. Ammonium carbonate solution (5% w/v) was pre-pared by adding 50.0 g of $(NH_4)_2CO_3$ to approximately 600 ml of water in a 1 liter volumetric flask, mixing until dissolved, warming to room temperature, and diluting to volume. The ammonium sulfate solution (5% w/v) consisted of nitric acid (J. T. Baker No. 5-9063; suitable for mercury determination) and sulfuric acid (J. T. Baker No. 5-9685; suita-ble for mercury determination). Purified deionized water was prepared by passing ordinary deionized water through two IWT Research Model I demineralizer cartridges (Arthur H. Thomas Co., No. 3923-D25) and then through a Puritan Model I cartridge (Arthur H. Thomas Co., No. 3923-D30). Necessary connections were made with Tygon tubing. This water was used for all procedures requiring water. All other chemicals were ACS reagent-grade quality.

All glassware used must be washed in 10% v/v nitric acid and rinsed with deionized water. The same glassware was used for all analyses and was not cleaned by any other method or mixed with other equipment.

The Fisher filtrator (Fisher Scientific Co., No. 9-788) used must be free of corrosion. A Büchner filter funnel with fritted disk was used (Kimble No. 28400-30M). The ends were drawn out to permit entry into a 10 ml volumetric flask. The resin tube was prepared by fusing a 24/40 TS female glass joint to the upper end of a 2.5 cm i.d. Allihn filter tube with a 145–175 μm glass frit (Ace Glass Co., No. 7195-02). The eluent reservoir consisted of a 500 ml graduated cylindrical separatory funnel supplied with Teflon stopcock plug and 24/40 TS male glass joint (Ace Glass Co., No. 7267-T). The flow rate was controlled with the stopcock. Absorption measurements were made with Perkin-Elmer Models 303 and 503 atomic absorption spectrometers equipped with deuterium background corrector, Model 056 recorder, and Perkin-Elmer Intensitron hollow-cathode lamps. Operating parameters were similar to those recommended by the manufacturer for flame determination.

The regenerated resin was placed in the filter tube to a depth of 1.0–1.2 cm (approximately 5 ml wet volume) and the separator attached. The column was equilibrated to pH 6.5 $\pm$ 0.5 by the addition of 1.0 ml of 1 N H_2SO_4 and 100 ml of pH 6.5 buffer. The sample was decanted into the separator and the flow rate controlled at 9 $\pm$ 2 ml/min. When the level of the liquid was just above the top of the resin bed, 200–300 ml of water and then 15 ml of 5% $(NH_4)_2SO_4$ were added. The effluent was discarded.

Procedure

Caution. Because of trace amounts of some metals of interest in the reagents used, equal amounts should be added to duplicates and reagent blank in all cases. Samples should be prepared in equipment that is free of contamination by the metals of interest.

Sample digestion. Weigh a 50.0 g sample into a 1500 ml beaker together with 20–30 mg of vanadium pentoxide, five or six glass boiling beads, and 5.0 ml of strontium solution. Carefully add 100 ml of concentrated nitric acid to each. Cover beakers with watch glasses. Heat mixtures on a hot plate with a gradual increase in temperature until the solutions reach a full boil. Any frothing that occurs can be controlled by the addition of small amounts of water. Heat mixtures until oxide of nitrogen fumes are no longer evolved. Cool solutions to room temperature and adjust the volume to about 200 ml with water. Add 20 ml of sulfuric acid to both sample and blank and evaporate until charring begins. Add hydrogen peroxide (50%) cautiously, dropwise, to each mixture until the sample solution is clear and the same green color as the reagent blank. Since 50% hydrogen peroxide is a very strong oxidant, add it slowly in small amounts to prevent frothing and splattering. The use of rubber gloves is recommended. For some commodities such as wheat, milk, and fish, it is advantageous to add additional nitric acid just after charring begins and then continue digestion.

Since the relatively large amounts of calcium in milk interfere with the co-precipitation of lead with strontium sulfate, 10.0 ml of strontium solution was added instead of 5.0 ml, and then the regular procedure was continued.

Place cooled digests in an ice bath. After several minutes rinse the sides down with three 15 ml portions of water (total 45 ml). Keep the beakers in the ice bath for at least 45 min and swirl occasionally. Fit the Büchner filter funnel with a rubber stopper and attach to the Fisher filtrator. Using a 400 ml beaker as a receiver, transfer the solution and precipitate to the funnel. Apply vacuum carefully and allow the solution to filter slowly. Turn off the vacuum and rinse the digest beaker into the funnel with 15 ml of chilled 1 N sulfuric acid. Reapply the vacuum. Repeat the rinse procedure twice, being sure the precipitate is transferred completely to the funnel. Use ammonium carbonate solution (5%, 10 ml) to rinse down the sides of the funnel. Apply gentle vacuum and filter the solution slowly. Add ammonium carbonate solution (30 ml) to the funnel and allow 25 min to elapse before applying vacuum. Repeat addition and filtration of ammonium carbonate solution.

Remove the beaker from the Fisher filtrator and save for the determination of cadmium, copper, manganese, nickel, and zinc.

Place a 10 ml volumetric flask in the filtrator under the tip of the funnel. Dissolve the precipitate with four separate 2.5 ml portions of warm 2 N nitric acid. Cool the mixture to room temperature and dilute to volume with 2 N nitric acid.

Determine lead at 2833 or 2170 Å against lead standards (0.25, 5.0, 7.5, and 10.0 μg/ml) that contain 5.0 ml of 2% strontium solution per 10.0 ml. Correct sample absorbances for the absorbance of the reagent blank.

Add 3 drops of 0.05% Methyl Red to the filtrate from the lead determination. Agitate the mixture carefully to offset frothing and spattering caused by neutralization of the carbonate and place in an ice bath. Adjust the pH to yellow by careful addition of 50% sodium hydroxide. Cool the mixture to room temperature and use a pH meter to complete adjustment to pH 6.5 ± 0.5 with 1 N sodium hydroxide and/or 1 N sulfuric acid.

Pour the filtrate through the column and wash with 5% ammonium sulfate as described under the section on column preparation. Elute with 2 N sulfuric acid, collecting the first 23–24 ml of eluent in a 25 ml volumetric flask. Dilute to volume with 2 N sulfuric acid. The resulting solution is approximately 1.6 N in sulfuric acid due to dilution by one bed volume of ammonium sulfate solution. Determine the absorbance for each metal vs. freshly prepared standards and read the concentration from a standard curve or calculate from the absorbance of the nearest standard. Correct each standard sample reading for the standard blank (zero concentration) and for the reagent blank. For metals such as zinc that are often present in high concentrations, serial dilutions of an aliquot of the sample with 1.6 N sulfuric acid are prepared to bring the concentration into the measurable range.

Determination of Iron, Copper, Nickel, and Manganese in Fats and Oils

See the procedure on p. 275. This can be readily used for food products.

5
Analysis of Metals and Alloys

Atomic absorption has been applied to the analysis of both major and trace constituents of alloys. Stringent requirements of precision and accuracy, often demanded for major elements, may negate the use of atomic absorption for these components.

Atomic spectral interference is rarely encountered in atomic absorption work. However, perhaps its highest probability exists in the analysis of high-purity metals.

Larkins and Willis (141) cite an important example. In the determination of zinc in high-purity copper metal, overlap of the 2138.56 Å zinc resonance line with the 2138.53 Å low-lying excited-state copper line occurs. Using a 20,000 μg/ml copper solution, free of zinc, a signal proportional to 5 μg/ml zinc is recorded. When a problem of this type is encountered, the only solution is to extract the analyte out of the interfering matrix.

SAMPLING

Although appearing homogeneous metals are usually quite the opposite. Trace constituents are frequently concentrated in blobs throughout the sample, in a thin layer on the surface, or at the bottom of the sample. It is important, therefore, in sampling alloys or high-purity metals to take several portions over the total volume. This can be done by drilling out sample turnings at various positions in the sample.

DISSOLUTION

Depending on alloy composition, a variety of acids or fluxes are employed for dissolution. Acids commonly used include hydrochloric,

220

nitric, sulfuric, perchloric, and/or combinations of these. For samples containing zirconium, titanium, etc., which tend to readily form oxides, hydrofluoric acid is useful. Again, as with other sample types, the use of sulfuric and hydrochloric acids invalidates results for lead and silver, respectively.

It is preferable to use acids for dissolution where possible. Fluxes such as sodium peroxide, sodium hydroxide, and potassium hydroxide may be required for special alloys.

It is often not possible to analyze all the desired elements using a single decomposition. For example, if nitric acid is used for dissolution of an alloy, tin and antimony may not be soluble. Likewise, when using hydrochloric acid as the solvent, silver and sometimes lead will not be soluble.

STANDARD SOLUTIONS

To overcome interference problems often encountered from the predominant matrix metal in an alloy or high-purity metal, it may be necessary to add this element to standard solutions. It is important to obtain ultrahigh-purity metal for this purpose. Spec-pure and 99.999% pure metals may not be sufficiently free of the analyte elements.

Determination of a Number of Elements in a Wide Range of Steels (142)

Comment on Method

The following elements can be determined by the method: manganese, nickel, chromium, molybdenum, copper, vanadium, cobalt, titanium, tin, aluminum, and lead. A single dissolution in hydrochloric and nitric acids, taken to fumes in perchloric acid, is used. Iron must be added to standards in the amount present in samples. A number of BCS steel standards are analyzed and acceptable agreement was obtained by the procedure.

Reagents—Stock Solutions

Manganese (1000 mg/liter). Dissolve 1.0000 g of pure manganese metal in 50 ml of hydrochloric acid (sp. gr. 1.18) and dilute to 1 liter.

Nickel (1000 mg/liter). Dissolve 1.0000 g of pure nickel metal in 40 ml of 6 N nitric acid and dilute to 1 liter.

Chromium (1000 mg/liter). Dissolve 1.0000 g of pure chromium metal in 30 ml hydrochloric acid (sp. gr. 1.18) and dilute to 1 liter.

Molybdenum (1000 mg/liter). Dissolve 1.829 g of analytical reagent grade ammonium molybdate in water and dilute to 1 liter.

Copper (1000 mg/liter). Dissolve 1.0000 g of pure copper metal in 50 ml of 6 N nitric acid and dilute to 1 liter.

Vanadium (1000 mg/liter). Dissolve 2.296 g of analytical reagent-grade ammonium vanadate in 20 ml of 100 volume hydrogen peroxide and dilute to 1 liter.

Cobalt (1000 mg/liter). Dissolve 1.0000 g of pure cobalt metal in 50 ml of 6 N nitric acid and dilute to 1 liter.

Titanium (1000 mg/liter). Dissolve 7.394 g of analytical reagent grade potassium titanium oxalate in water and dilute to 1 liter.

Tin (1000 mg/liter). Dissolve 1.0000 g of pure tin metal in 50 ml of hydrochloric acid (sp. gr. 1.18) plus 5 ml of nitric acid (sp. gr. 1.42). Add 150 ml of hydrochloric acid (sp. gr. 1.18) and dilute the solution to 1 liter.

Aluminum (1000 mg/liter). Dissolve 1.0000 g of pure aluminum metal in 25 ml of hydrochloric acid (sp. gr. 1.18) and add a few drops of nitric acid (sp. gr. 1.42). Dilute the solution to 1 liter.

Lead (1000 mg/liter). Dissolve 1.0000 g of pure lead metal in 10 ml of 2 N nitric acid and dilute to 1 liter.

Tungsten (1000 mg/liter). Dissolve 1.420 g of ammonium tungstate in water and dilute to 1 liter.

Dilute solutions, when needed, were prepared by diluting the above concentrated solutions, and should be prepared daily.

Iron (5%). Dissolve 5 g of high-purity iron (BCS 260/3) in 40 ml of hydrochloric acid (sp. gr. 1.18) plus 5 ml of nitric acid (sp. gr. 1.42). When the reaction is complete, add 20 ml of perchloric acid (sp. gr. 1.54) and evaporate the solution until fumes of perchloric acid just appear. Cool the sample and dilute to 100 ml with water.

Reagents—Perchloric Acid—Phosphoric Acid—Sulfuric Acid (Mixture A).

To 300 ml of water, add 100 ml of perchloric acid (sp. gr. 1.54), 100 ml of phosphoric acid (sp. gr. 1.75), and 100 ml of sulfuric acid (sp. gr. 1.84).

Procedure

Determination of manganese, nickel, chromium, molybdenum, copper, vanadium, cobalt, titanium, tin, aluminum, and lead. Weigh 1.0000 g of sample into a 250 ml beaker and dissolve in 10 ml of hydrochloric acid (sp. gr. 1.18). Add 5 ml of nitric acid (sp. gr. 1.42). After the initial reaction has subsided, add 10 ml of perchloric acid (sp. gr. 1.54) and evaporate the solution until the sample is fully oxidized and fumes of perchloric acid appear. (This is achieved when the solution turns red and perchloric acid is seen to reflux on the sides of the beaker.) For samples not containing chromium, the solution will not turn red but is merely allowed to fume for 5 min. Cool and dissolve the soluble salts in about 50 ml of water. Filter the solution through a Whatman No. 541 filter paper, wash well with water, and add the washings to the filtrate. Dilute the solution to 100 ml.

Preparation of calibration solutions. To each of seven 250 ml beakers, add 1.0 g of pure iron and suitable volumes of stock solutions (Table XXXVIII). The same dissolution procedure is used as specified above for the samples.

It is imperative that the chromium and molybdenum stock solutions be added before fuming takes place. However, if the chromium and molybdenum are to be omitted from the scheme, the stock solutions can be added after the iron has been dissolved for the calibration ranges. Alternatively, the pure iron stock solution can be added to the aqueous standards before dilution (i.e., 20 ml of pure iron stock solution in a final volume of 100 ml is equivalent to a 1% sample solution.

Determination of tungsten and of molybdenum when the concentration of tungsten is higher than 0.5%. Weigh 1.0000 g of sample into a 250 ml beaker. Add 50 ml of acid mixture A and heat gently. When the sample has dissolved, oxidize the solution by adding nitric acid (sp. gr. 1.42) dropwise. Evaporate the solution until the first fumes of perchloric acid appear. Cool the solution and dilute to 50 ml. Filter the solution through a Whatman No. 541 filter paper and dilute to 100 ml.

Table XXXVIII

Calibration Solutions[a]

Manganese							
Dilute stock solution (100 mg/l)/ml	0	1.0	2.0	5.0	10.0	15.0	20.0
Concentration (mg/l)	0	1	2	5	10	15	20
Manganese (%)	0	0.01	0.02	0.05	0.10	0.15	0.20
Nickel							
Stock solution (200 mg/l)/ml	0	1.0	2.5	5.0	10.0	15.0	20.0
Concentration (mg/l)	0	2	5	10	20	30	40
Nickel (%)	0	0.02	0.05	0.10	0.20	0.30	0.40
Chromium (up to 0.5%)							
Stock solution (500 mg/l)/ml	0	1.0	2.0	4.0	6.0	8.0	10.0
Concentration (mg/l)	0	5	10	20	30	40	50
Chromium (%)	0	0.05	0.1	0.2	0.3	0.4	0.5
Chromium (up to 4.0%)							
Stock solution (1000 mg/l)/ml	0	1.0	2.0	4.0	6.0	8.0	10.0
Concentration (mg/l)	0	10	20	40	60	80	100
Chromium (%) (0.25-g sample)	0	0.4	0.8	1.6	2.4	3.2	4.0
Chromium (up to 12.0%)[b]							
Stock solution (100 mg/l)/ml	0	2.5	5.0	7.5	10.0	20	30
Concentration (mg/l)	0	25	50	75	100	200	300
Chromium (%) (0.25-g sample)	0	1	2	3	4	8	12
Molybdenum (up to 0.2%)							
Stock solution (100 mg/l)/ml	0	2.0	5.0	10.0	15.0	20.0	
Concentration (mg/l)	0	2	5	10	15	20	
Molybdenum (%)	0	0.02	0.05	0.10	0.15	0.20	
Molybdenum (up to 0.5%)							
Stock solution (500 mg/l)/ml	0	1.0	2.0	4.0	6.0	8.0	10.0
Concentration (mg/l)	0	5	10	20	30	40	50
Molybdenum (%)	0	0.05	0.1	0.2	0.3	0.4	0.5
Molybdenum (up to 1.0%)							
Stock solution (100 mg/l)/ml	0	1.0	2.0	4.0	6.0	8.0	10.0
Concentration (mg/l)	0	10	20	40	60	80	100
Molybdenum (%)	0	0.1	0.2	0.4	0.6	0.8	1.0

Copper							
Dilute stock solution (100 mg/l)/ml	0	1.0	2.0	5.0	10.0	15.0	20.0
Concentration (mg/l)	0	1	2	5	10	15	20
Copper (%)	0	0.01	0.02	0.05	0.10	0.15	0.20
Vanadium							
Stock solution (1000 mg/l)/ml	0	1.0	2.0	4.0	6.0	8.0	10.0
Concentration (mg/l)	0	10	20	40	60	80	100
Vanadium (%)	0	0.1	0.2	0.4	0.6	0.8	1.0
Cobalt							
Stock solution (500 mg/l)/ml	0	1.0	2.5	5.0	10.0	15.0	20.0
Concentration (mg/l)	0	5	10	20	30	40	50
Cobalt (%)	0	0.05	0.10	0.20	0.30	0.40	0.50
Titanium							
Stock solution (1000 mg/l)/ml	0	1.0	2.5	5.0	10.0	15.0	20.0
Concentration (mg/l)	0	10	25	50	100	150	200
Titanium (%)	0	0.10	0.25	0.50	1.0	1.5	2.0
Tin							
Stock solution (500 mg/l)/ml	0	1.0	2.0	4.0	6.0	8.0	10.0
Concentration (mg/l)	0	5	10	20	30	40	50
Tin (%)	0	0.05	0.10	0.20	0.30	0.40	0.50
Aluminum							
Stock solution (500 mg/l)/ml	0	1.0	2.0	4.0	6.0	8.0	10.0
Concentration (mg/l)	0	5	10	20	30	40	50
Aluminum (%)	0	0.05	0.10	0.20	0.30	0.40	0.50
Lead							
Dilute stock solution (100 mg/l)/ml	0	1.0	2.0	5.0	10.0	10.0	20.0
Concentration (mg/l)	0	1	2	5	10	15	20
Lead (%)	0	0.01	0.02	0.05	0.15	0.15	0.20
Tungsten[c]							
Stock solution (1000 mg/l)/ml	0	5.0	10.0	20.0	30.0	40.0	
Concentration (mg/l)	0	50	100	200	300	400	
Tungsten (%)	0	0.5	1	2	3	4	

[a] These are based on a 1.0-g sample (except when otherwise stated), and a final volume of 100 ml is used throughout.
[b] For this range the burner must be in the fully rotated position.
[c] 50 ml of acid mixture A must be added to each of the tungsten calibration solutions (tungsten higher than 0.5%).

Preparation of calibration solution.   To each of seven 250 ml beakers add 1.0 g of pure iron and suitable volumes of stock solutions. Use the dissolution procedure specified above for the samples.

Extension of calibration ranges.   When it is required to determine higher concentrations of alloying elements than those provided for by the recommended calibration ranges, the sample solution should be diluted by an appropriate factor such that the final solution still contains 1% iron. All observations of freedom from interelemental interferences were made on solutions containing 1% iron. It may therefore be taken as a general principle that, provided both samples and standards contain 1% iron, it is possible to use any degree of dilution.

In preparing samples known to require range extension, the appropriate amount of 5% stock iron solution can be added before finally making up to volume. A sample already made up to volume can simply be diluted as required with 1% iron solution derived by appropriate dilution of the 5% stock solution.

Analysis.   Recommended instrumental conditions for common constituent elements are summarized in Table XXXIX.

The blank ("zero" standard) and calibration solutions should be aspirated followed by the sample solutions. For highest accuracy the blank solution should be run between all other standards and samples and a high standard repeated after every five or ten samples. The calibration graph of absorbance vs. concentration is plotted for each element and the concentrations of the elements in the sample solutions are read off.

Determination of Arsenic, Antimony, and Tin in Steels (143)

Comment on Method

The method is for steel samples containing 0.0020–1% arsenic antimony, and tin. Electrothermal atomization is used and the detection limits for the elements were found to be arsenic: 1×10^{-9} g; antimony; 3×10^{-10} g; tin: 8×10^{-10} g. The dissolution must be made in nitric acid since other acids cause serious signal loss in the case of arsenic and tin.

Reagents and Equipment

A Varian-Techtron AA5 atomic absorption spectrometer was used in conjunction with a Varian-Techtron Model 63 carbon rod atomizer. This has a carbon furnace in the form of a cylindrical graphite tube 3 mm in diameter and 9 mm in length. A solution of the sample is injected

Table XXXIX

Instrumental Conditions

	Mn	Ni	Cr[c]	Mo	Cu	V	Co	Ti	Sn	Al	Pb	W
Wavelength (Å)	2795	2320	3579	3133	3248	3184	2407	3643	2240	3093	2170	2551
Slit width (mm)	0.03	0.05	0.05	0.05	0.05	0.05	0.03	0.05	0.10	0.05	0.05	0.05
Burner[a]	A	A	N	N	A	N	A	N	N	N	A	N
Observation height (cm)	0.8	0.8	0.5	0.5	0.8	0.7	0.8	0.7	0.7	0.7	0.8	0.8
Air (l/min)	5.0	5.0	—	—	5.0	—	5.0	—	—	—	5.0	—
Acetylene (l/min)[b]	1.4	0.8	4.2	4.7[d]	1.0	4.5	1.2	4.5	4.4	4.2	1.2	4.5
Nitrous oxide (l/min)[b]	—	—	5.0	5.0	—	5.0	0	5.0	5.0	5.0	—	5.0

[a] A is a 1-cm air-acetylene burner, and N a 5-cm nitrous oxide-acetylene burner with 0.59-mm jaw width (see Apparatus).

[b] These values should be used as a guide only and the flow-rate must be adjusted to give the best sensitivity for each element.

[c] For the chromium range 25-300 1/min, the burner must be in the fully rotated position to reduce the sensitivity (for the SP90 atomic absorption spectrophotometer this is about 50° from the optical axis).

[d] This acetylene flow-rate is critical and must be carefully adjusted to give a flame with a maximum red "feather" height without luminescence.

through a small hole in the side of the tube and the atoms produced by electrically heating the furnace are viewed along the tube axis. A Hamamatsu TV Type R106 photomultiplier was used in place of the standard HTVR 213 tube and the signals produced were recorded on a Honeywell Electronik 19 recorder. Argon was used as the inert shield gas at a flow rate of 4 liters/min. Hollow-cathode lamps (Varian-Techtron) were used as the radiation sources for arsenic, antimony, and tin. The lamps were run at currents 7, 10, and 8 mA, respectively, and the wavelengths used were 1937, 2176, and 2246 Å, respectively. A microliter syringe (Varian-Techtron) was used to inject the samples into the carbon tube furnace.

Deionized water was used throughout.

British Chemical Standard Steel Nos. 320-330 and 260/3 containing known amounts of arsenic, antimony, and tin were dissolved in dilute nitric acid as described below.

Procedure

Weigh accurately 1.0 g of steel, in the form of millings or drillings, into a 250 ml borosilicate beaker and add 30 ml of dilute (A.R.) nitric acid (1:4). Warm gently until dissolution is complete. Transfer the solution to a 100 ml volumetric flask and make up to the mark with water. Smaller steel samples can be similarly prepared by keeping the sample weight, volume of dilute nitric acid, and final solution volume in the same ratio.

Mix the sample solution well, and determine the arsenic, antimony, and tin by injecting successive 5 μl volumes into the carbon furnace and comparing the signals produced with those from the similarly prepared calibration solutions.

The most suitable instrument operating conditions found for measuring these three elements are given in Table XL.

Determination of Microgram Amounts of Antimony, Bismuth, Lead, and Tin in Aluminum, Iron, and Nickel-Based Alloys (144)

Comment on the Method

The analyte metals are extracted from the nickel matrix using a nonaqueous system of trioctylphosphine oxide (TOPO) in MIBK. Iron(III) commonly produced during decomposition of the sample is reduced to iron(II) using ascorbic acid. After this precaution has been taken little interference from any of the constituents of these alloys is encountered. Copper might be expected to be the greatest problem.

Table XL

Optimal Operating Conditions for Determination of Arsenic, Antimony, and Tin

Element	As	Sb	Sn
Wavelength setting (Å)	1937	2176	2246
Slit width setting (μm)	200	200	200
Sample volume (μl)	5	5	5
Lamp current (mA)	7	10	8
Furnace conditions[a]			
Dry V (sec)	3(30)	3(30)	3(30)
Ash V (sec)	4(15)	4(15)	4(15)
Step atomize V (sec)	6(6)	6.5(5)	—
Ramp atomize V (sec)	—	—	7(5)
Recorder sensitivity			
(mV fsd)	2	20	5

[a] For each step, the voltage is given followed by the time (in parentheses). In all cases, the argon flow was 66 ml/sec.

However, tests show that 100 μg of antimony can be determined in the presence of up to 50 mg of copper with no interference. Another advantage of the procedure is that a close control of extraction variables is not required.

The procedure was tested on a variety of standard alloys. Satisfactory agreement with accepted values was obtained.

Reagents and Equipment

A Perkin-Elmer Model 303 atomic absorption spectrometer, equipped with a premix chamber and triple-slot Boling and nitrous oxide heads, was used. Perkin-Elmer hollow-cathode lamps were used as light sources lamps and a Sargent Model SRG recorder was employed as a readout device. The instrument settings are summarized in Table XLI.

Nonaqueous standards. A stock solution containing 1000 μg/ml of antimony, bismuth, lead, and tin is prepared as follows (see Table XLII). Dissolve 0.4367 g of dibutyltin bis(2-ethylhexanoate), 0.2632 g of lead cyclohexanebutyrate, 9.2585 g of triphenyl bismuth, and 0.2900 g of triphenylstibine in 10 ml of 2-ethylhexanoic acid plus 20 ml of MIBK. Dilute the sample to 100 ml with MIBK. Prepare a dilute solution from the stock solutions by dilution with MIBK to contain 50 μg/ml of the test elements. A precipitate appears 2 weeks after preparation of the stock and dilute solutions if 2-ethylhexanoic acid was not present.

Table XLI

Operating Conditions for Perkin-Elmer 303 Atomic Absorption Spectrometer When Aspirating MIBK Solutions

Element	Wave length (Å)	Slit width (Å)	Lamp current (mA)	Working range (μg ml^{-1})	C_2H_2 at 8 psig	Gas flow rates (min^{-1}) at 30 psig	
						Air	N_2O
Antimony	2176	2	35	1 to 20	2	24	—
Bismuth	2231	2	30	1 to 10	2	24	—
Lead	2170	7	30	1 to 7	2	24	—
Tin	2863	7	30	1 to 50	2	—	12

Standard stock solutions (1000 μg/ml). A stock solution is prepared from high-purity metal containing all the elements of interest. It should be 6 M in hydrochloric acid to prevent hydrolysis.

TOPO (5% solution in MIBK). 12.5 g of TOPO (Eastman 7440) is dissolved in 250 ml of MIBK.

Iodide reagent. A solution containing 30% w/v of potassium iodide and 10% w/v of ascorbic acid in 10% w/v hydrochloric acid is prepared daily.

All chemicals used were of analytical reagent grade. Grade A calibrated glassware was used.

Procedure

Weigh samples that contain a sufficient amount of antimony, bismuth, lead, and tin to give a satisfactory signal accurately into beakers. Instrumental response can be varied by adjusting the sample weight and volume of extractant as shown by the following example for the 10 μg/ml level:

Sample weight (g):	1	1	5	5
Final Volume (ml):	10	5	5	2
Element (μg/ml):	1	5	10	25

Dissolve the sample and remove any oxidant used in the sample preparation (see Table XLIII). Rinse the sides of the beaker with about 5 ml of water and add 4 g of ascorbic acid to those samples that contain iron(III) chloride to judge the completeness of the reduction to iron(II). Add 15 ml of the iodide reagent and transfer the sample to a 150 ml separating funnel, rinsing the beaker with water. Adjust the final volume to about 50 ml with a pipet. Introduce 10 ml of the 5% TOPO-MIBK reagent into the separating funnel, equilibrate the solutions for 30 sec,

Table XLII

Compounds Used to Prepare a Nonaqueous Standard Stock Solution

Organometallic compound	Molecular weight	Metal, % (certified)
Dibutyltin bis(2-ethylhexanoate)	519.3	Sn 22.9
Lead cyclohexanebutyrate	545.7	Pb 38.0
Triphenylbismuth	440.3	Bi 38.7[a]
Triphenylstibine	353.1	Sb 34.8[a]

[a] Metal content calculated for pure compound.

and allow phases to separate. Drain the aqueous layer off and discard. Transfer the organic phase into a 15 ml stoppered vial. Occasionally an emulsion forms, which can be broken by briefly centrifuging the organic phase. Phase-separation paper must not be used to break the emulsion unless it has been found to be free from a tin compound that is soluble in MIBK.

Optimize the burner height, aspiration rate, and gas flow rates; the conditions given in Table XLI can be used as guidelines. Prepare calibration solutions by diluting the standard stock solution with MIBK. Beginning with the solution that has the lowest metal content, aspirate each solution and record its absorbance. Aspirate the sample solutions and blank and record their absorbances. Correct the absorbance readings of the samples for any blank.

Determination of Lead, Silver, Zinc, Bismuth, and Cadmium in High Nickel Alloys (145)

Comment on the Method

The trace elements are separated from the nickel matrix by ion exchange. Cation ion exchange is used for the separation of the trace elements from the nickel matrix (Fig. 38). Three solutions, containing the elements of interest are obtained by elution, i.e., zinc solution, bismuth–cadmium solution, and a solution containing lead and silver along with titanium and molybdenum impurities. The latter is passed through an anion exchange resin to separate lead and silver from the impurities (Fig. 39).

Flame atomic absorption is used for the determination of all of the elements. The efficiency of recovery of the procedure was tested using nickel pellets to which known amounts of trace elements were added and by analyzing NBS SRM 361. Recoveries ranged from 92 to 100%.

Table XLIII

Sample Preparation Scheme for Solvent Extraction

Matrix	Solvent (ml of acid)			Final preparation for 1-g sample (graphite and silica removed by filtration)
	1 g	5 g	10 g	
Al	$HCl(15)$ + 30% H_2O_2[a]	$HCl(30)$ + 30% H_2O_2	$HCl(60)$ + 30% H_2O_2	Evaporate to 5 ml of HCl
Fe	$HCl(15)$ + $HNO_3(5)$	$HCl(30)$ + HNO_3 (10)	$HCl(60)$ + $HNO_3(20)$	Evaporate to 5 ml of HCl 10 ml of HCl + heat + HCOOH;[a] evaporate nearly to dryness and add 5 ml of HCl
Ni	$HNO_3(10)$ + 10 ml of H_2O	$HNO_3(30)$ + 30 ml of H_2O	$HNO_3(60)$ + 60 ml of H_2O	10 ml of HCl + heat + HCOOH; evaporate nearly to dryness and add 5 ml of HCl
Nickel oxide	$HCL(15)$	$HCl(30)$	$HCl(60)$	Reduce volume to 5 ml
Ni-Cr	$HCl(15)$ + $HNO_3(15)$	$HCl(30)$ + $HNO_3(10)$	$HCl(60)$ + $HNO_3(20)$	10 ml of HCl + heat + HCOOH; evaporate nearly to dryness and add 5 ml of HCl

[a] H_2O_2 and HCOOH are added dropwise, the latter until brown fumes due to the oxides of nitrogen disappear.

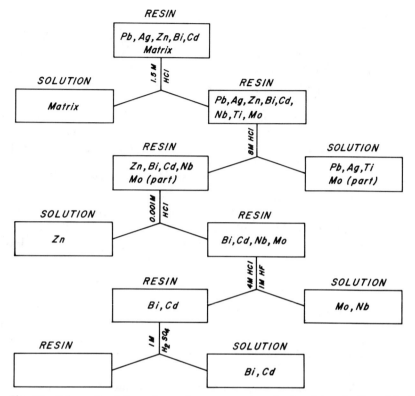

Fig. 38. Schematic of the anion exchange procedure for the determination of lead, silver, zinc, bismuth, and cadmium in high-nickel alloys (145).

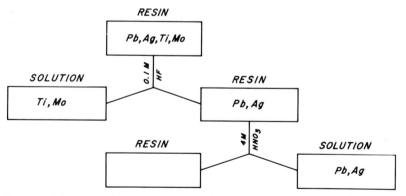

Fig. 39. Schematics of the anion exchange procedure for the separation of lead and silver from molybdenum and titanium impurities (145).

Reagents and Equipment

Reagent-grade chemicals were used in preparation of all standards and eluants. Stock solutions (1000 μg/ml) of lead, silver, bismuth, or cadmium were made with either the pure (99.999%) metal or commercially prepared solutions (J. T. Baker, "Dilute-It") that contained the elements in substrate form.

Anion exchange columns contained BioRex 9 (100–200 mesh, chloride form) resin in a plexiglass column. The resin bed was 14-cm-deep and 2.5-cm-in.-diameter. The column was preconditioned with 100 ml of 1.5 M hydrochloric acid.

Cation exchange columns contained BioRad AG50-X8 (100–200 mesh, H$^+$ form resin). The resin bed was 7-cm-deep and 2.5-cm-in.-diameter. The column was preconditioned with 100 ml of 0.1 M hydrofluoric acid.

A Perkin-Elmer Model 303 atomic absorption spectrometer was used. It was equipped with a three-slot burner and Perkin-Elmer hollow-cathode source lamps.

Two sets of calibration standards were prepared for each element. One set was a low standard for the 10× scale and the other a high standard for the 1× and 2× scales. The lead and bismuth contents were 1, 2, and 3 μg/ml for the low standards and 5, 10, and 15 μg/ml for the high standards. The zinc and cadmium contents were 0.25, 0.50, and 0.75 μg/ml for the low standards. Silver contents were 0.1, 0.2, and 0.3 μg/ml for the low standards and 1, 2, and 3 μg/ml for the high standards. Each calibration standard contained the same amount and type of acid as the final solution to be measured.

Ion Exchange Procedure

Figure 38 illustrates the procedure used to separate the lead, silver, zinc, bismuth, and cadmium from the alloy matrix.

Dissolve a 10 g sample of the alloy in appropriate acids, usually a mixture of hydrochloric and nitric acids. After the reaction ceases, add 1 ml of hydrofluoric acid (40%) and evaporate the solution to dryness. Add 20 ml of 12 M hydrochloric acid and evaporate to dryness.

Repeat the addition and evaporation of 12 M hydrochloric acid until all nitric acid is expelled. To ensure that all the salts are soluble, dissolve the residue in 50 ml of 12 M hydrochloric acid and evaporate to dryness on a steam bath. To produce the 1.5 M hydrofluoric acid solution required for the column, add 5 ml of hydrofluoric acid (40%) and 100 ml of water, and heat to redissolve the salts. Transfer the hydrochloric–hydrofluoric acid solution of the sample to the anion exchange column, and elute most of the matrix with 100 ml of 1.5 M HCl.

Elute the lead and silver from the column with 200 ml of 8 M hydrochloric acid; this fraction will also contain part of any titanium and molybdenum present in the sample, but these are later separated from the lead and silver by cation exchange. Elute the zinc with 200 ml of 0.001 M hydrochloric acid. Remove any organic compounds present in this eluate by addition of 5 ml of nitric acid and evaporation to dryness. At this point in the elution sequence elute any remaining molybdenum or any niobium that may have been partially absorbed on the column with 300 ml of 4 M hydrochloric–1 M hydrofluoric acid solution; discard the eluate.

Elute the bismuth and cadmium with 300 ml of 1 M sulfuric acid. Remove any organics in the eluate by adding 5 ml of 16 M nitric acid and evaporating to fumes of sulfur trioxide. Then cool, transfer to a glass beaker, and evaporate to dryness.

Remove titanium and molybdenum from the lead–silver fraction by the cation exchange procedure shown in Fig. 39. First remove any organics present in the eluate by addition of 20 ml of 16 M nitric acid and evaporation to dryness.

Convert the metals to fluorides by addition of 10 ml of 40% hydrofluoric acid, and evaporation to dryness. Produce the 0.1 M hydrofluoric acid solution required for the column by adding 5 ml of water and 2 ml of 40% hydrofluoric acid, redissolving the salts by heating and diluting with 300 ml of water. Transfer the solution to the cation exchange column, and elute the titanium and molybdenum with 300 ml of 4 M nitric acid. Obtain the chloride medium needed for analysis by addition and evaporation of 12 M hydrochloric acid until all nitric acid is removed from the eluate.

The separated and concentrated elements in the three (lead–silver, zinc, and bismuth–cadmium) elution fractions are now ready for analysis by atomic absorption. To determine quantities of the trace elements that may have been introduced by the reagents, take a reagent blank through the entire procedure with the same amounts of reagents used for the sample and deduct the reagent blanks from the sample values.

The sizes of volumetric flasks used are listed in Table XLIV. If concentration ranges other than those listed must be used, dilute the samples accordingly. A chloride medium is needed for lead and silver and the lead–silver fraction is transferred to flasks with 5% (v/v) hydrochloric acid. Zinc, bismuth, and cadmium require a nitrate medium, and those fractions are transferred with 5% (v/v) nitric acid. The instrumental parameters used are listed in Table XLIV; in general, the instrument settings correspond to those recommended in the analytical methods

Table XLIV

Flask Sizes and Concentration Ranges for 10-g Samples and Instrumental Parameters

Element	Flask size (ml)	Conc range ppm	Scale expansion	Wavelength (Å)	Slit
Lead[a]	5	<1.5	10×	2833	4[c]
		1.5–7.5	2×		
Silver[b]	5	<0.15	10×	3281	4
		0.15–1.5	1×		
Zinc[a]	25	<1.9	2×	2139	4
		1.9–3.8	1×		
Bismuth[a]	5	<1.5	10×	2231	3
		1.5–7.5	2×		
Cadmium[b]	5	<0.15	10×	2288	4
		0.15–0.38			

[a] Air–acetylene flame.
[b] Air–propane flame.
[c] Slit 4 corresponds to 7 Å, and 3 to 2 Å.

supplied with the unit. Obtain an absorbance value for each element, and establish the concentrations from calibration curves.

Determination of Antimony, Arsenic, Bismuth, Cadmium, Lead, and Tin in Iron, Copper, and Zinc Alloys (146)

Comment on the Method

A graphite furnace atomizer is used in this method. Antimony and lead can be determined by using standards containing the main component of the alloy as a matrix ingredient. Bismuth, cadmium, and tin require use of the standard addition method for the copper alloy. In the case of iron alloys, matrix standards were satisfactory.

Samples are dissolved in nitric acid, which is satisfactory for all cases except for tin in copper alloys. Suprisingly, nitric acid could be employed for tin in zinc alloys. Table XLV shows the alloys analyzed and the elements determined.

Reagents and Equipment

All analyses were done with a Perkin-Elmer Model HGA 2100 graphite furnace, with either a Perkin-Elmer Model 503 or 305B atomic absorption spectrometer. Both instruments were equipped with simul-

Table XLV

Alloys Analyzed

Matrix	Source	Elements determined
Mild Steel	NBS	As, Pb, Sb
	British chemical standards	
Copper base alloys	British nonferrous metals research association[a]	As, Pb, Sn, Sb, Bi
	NBS	
Zinc alloys	Zinc et Alliages[a]	Pb, Sn, Cd
	NBS	

[a] Obtained through Brammer Standard Company.

taneous deuterium arc background correctors, which were used for all determinations. All data were obtained by reading peak heights, either on a recorder or with the peak reader of the Model 503.

The purge gas for all analyses was nitrogen. Eppendorf micropipets were used to introduce 20 μl of the solutions (samples or standards) into the graphite furnace. Perkin-Elmer electrodeless discharge lamps were used for antimony, arsenic, cadmium, and tin determinations. An Intensitron hollow-cathode lamp was used for bismuth. Lead was determined in steel with an electrodeless discharge lamp, while an Intensitron lead hollow-cathode lamp was used for the zinc and copper alloys.

Procedure

Dissolve all alloys (0.5 g samples) by adding 10 ml of 1:1 nitric acid (Ultrex grade, J. T. Baker Co.), and heating if necessary. Dilute the samples to 50 ml. Dilute these stock solutions further with water as required to bring the concentrations of the elements of interest within the optimal analytical range. Allow undissolved residues from samples containing carbon or silica to settle before taking an aliquot for further dilution.

Prepare standards by dilution of commercial atomic absorption standards. To each standard add the principal alloy component as a synthetic matrix. For example, when brass samples are analyzed, all standards should contain an amount of pure copper equivalent (on a weight basis) to the concentration of copper in the samples.

The instrument operating conditions are shown in Table XLVI.

Arsenic. Dilute the samples to give matrix concentration of 0.02 or 0.05%; this keeps all arsenic absorbance values below 0.3, where the calibration curve is linear. Use the standard addition method for the

Table XLVI

Operating Conditions

	As	Bi	Cd	Pb	Sb	Sn
Wavelength (Å)	1937	2231	2288	2833	2176	2863
Drying temp (°C)	125	125	100	125	100	100
Charring temp (°C)	500	500	250	500	500	1000
Atomization temp (°C)	2700	2500	2100	2500	27000	2700
Nitrogen purge flow (units)	30	30	50	30	30	Interrupt

copper alloy but the matrix standard approach is satisfactory for the iron alloy.

Atinomy and lead.   Use matrix standards containing a similar concentration of the principal matrix ingredient.

Cadmium and bismuth.   Use the standard addition method. In the case of cadmium, large dilutions are often required to bring it into the working range.

Tin.   Tin can be analyzed by the standard addition method in zinc alloys after a nitric acid dissolution. However, the copper alloy cannot be done in this manner. Even use of hydrochloric acid as the decomposing acid does not solve this problem.

Table XLVII shows the detection limits for the elements.

Table XLVII

Detection Limits in Alloys

	Detection limits		
Element	HGA (pg) aqueous	HGA(μg/g)[a] metal matrix	Flame (μg/g)[b,c] metal matrix
As	10	0.2	17
Bi	10	0.2	16
Cd	0.3	0.005	0.2
Pb	5	0.08	4
Sb	20	0.3	9
Sn	10	0.2	7

[a] Assumes 20 μl of a 1% (w/v) solution of the metal matrix.
[b] Assumes a 1% (s/v) solution of the metal matrix.
[c] 1 μg/g = 0.0001 st %.

Traces of Lead in High-Purity Copper (147)

Comment on the Method

A method involving extraction or preconcentration of the lead is rejected because of possibilities of contamination and the time-consuming nature of such procedures. Instead a direct method for 1–1000 μg/ml lead in copper is proposed using an electrothermal atomizer. Copper, 99.999+%, is added to standards in amounts to match the samples to overcome signal depression effects. The method was tested on standard reference coppers, aluminum, brass, bronze, and phosphorous. Results by the proposed procedure were in good agreement with certificate values.

Reagents and Equipment

Reagents of the highest available purity were used and solutions were prepared with distilled water from a quartz still.

Stock solution of lead (100 μg/ml lead).   Dissolve 0.16 g of analytical reagent-grade lead nitrate in water, transfer the solution to a 1 liter calibrated flask and dilute to the mark with water and sufficient AnalaR nitric acid to make the final solution 0.01 M in nitric acid.

Stock solution of copper (10,000 μg/ml copper).   Dissolve 0.5 g high-purity copper (from American Smelting and Refining Co., ASARCO-A-58 grade 99.999% copper) in 10 ml of 40% AnalaR nitric acid, transfer to a 50 ml calibrated flask, and dilute to the mark with water.

Stock solution of zinc (3000 μg/ml zinc).   Dissolve 0.15 g of zinc metal in 10 ml of 40% AnalaR nitric acid, transfer to a 50 ml calibrated flask, and dilute to the mark with water.

The instrument used for all measurements was a Perkin-Elmer Model 306 atomic absorption spectrometer equipped with an HGA 70 heated-graphite atomizer and a deuterium arc background corrector and coupled to an Electronik 19 strip-chart recorder. A Perkin-Elmer Intensitron hollow-cathode lamp was used as the source, with a lamp current of 8 mA. The lead wavelength of 2833 Å was used with a spectral band of 7 Å. Samples were transferred to the center of the tube by means of 50 or 20 μl Eppendorf pipet. In all cases the intermediate charring or ashing cycle was omitted. The argon flow rate was 1.5 liter/min at 10 psi.

Procedures

Procedural detail will depend to some extent on the desired concentration range. No changes are necessary in going from determinations of lead in high-purity copper to those in copper-based alloys.

Samples Containing 0.0001–0.0010% Lead

Preparation of calibration solutions. Dilute 10 ml of stock lead solution (100 μg/ml) to 1 liter with water. This solution should be freshly prepared every day. Transfer 0, 1.0, 2.0, 3.0, 4.0, and 5.0 ml of this solution into 100 ml Teflon beakers each containing 0.5 g ASARCO-A-58 grade copper metal and add 10 ml of 40% nitric acid. When the copper has dissolved, transfer the solutions to 50 ml calibrated flasks and dilute to the mark with water. These solutions contain the equivalent of 0, 0.0002, 0.0004, 0.0006, 0.0008, and 0.0019% lead in copper when 0.5 g of sample is used to prepare 50 ml of solution.

Preparation of sample solutions. Weight 0.5 g of sample into a 100 ml Teflon beaker and dissolve in 10 ml of 40% HNO_3. Transfer the solution to a 50 ml calibrated flask and dilute to the mark with distilled water.

Sequentially inject 50 μl aliquots of standard and sample solutions into the graphite tube. Dry each for 40 sec at 100°C and atomize at 2100°C (7.5 V) for 10 sec. Record the absorption signals during atomization at 3× scale expansion. Interpolate sample concentrations from a calibration graph obtained from the standards.

Samples Containing 0.0010–0.0100% Lead

Preparation of calibration solutions. Dilute 10 ml of stock lead solution (100 μg/ml) to 100 ml with water. This solution should be freshly prepared every day. Transfer 0, 1.0, 2.0, 3.0, 4.0, and 5.0 ml of this solution to 100 ml Teflon beakers, each containing 0.5 ASARCO-A-58 grade copper, and add 10 ml of 40% nitric acid. When the copper has dissolved, transfer the solutions to 50 ml calibrated flasks and dilute to the mark with water. These solutions contain the equivalent of 0, 0.0020, 0.0040, 0.0060, 0.0080, and 0.0100% lead in copper when 0.5 g copper sample is used to prepare 50 ml of solution.

Preparation of sample solutions. Proceed as in procedure above. Sequentially inject 20 μl aliquots of standard and sample solutions into the graphite tube. Dry each for 30 sec at 100°C and atomize at 2200°C (8.0 V) for 10 sec. Record the absorption signals during atomization with no scale expansion. Use a calibration graph to obtain sample concentrations.

Samples Containing 0.0100–0.100% Lead

Preparation of calibration solutions. Transfer 0, 1.0, 2.0, 3.0, 4.0, and 5.0 ml of the stock lead solution (100 ppm) to 100 ml Teflon beakers, each containing 0.5 g (May and Baker) copper turnings and add 10 ml of 40% nitric acid. When the copper has dissolved, transfer the solutions to 50 ml calibrated flasks and dilute to the mark with water. These solutions, when diluted by a factor of 10, contain the equivalent of 0, 0.020, 0.040, 0.060, 0.080, and 0.100% lead in copper when 0.5 g of copper sample is used to prepare 500 ml of solution.

Preparation of sample solutions. Proceed as in procedure above. Finally, dilute each solution by a factor of 10. Instrument operation is the same as in the 0.0010–0.0100% lead procedure above.

Rapid Method for the Dissolution of Lead Alloys and the Determination of Tin (148)

Comment on the Method

When lead is dissolved in hydrochloric acid, the reaction slows due to the deposition of insoluble chloride on the metal. Nitric acid can be used but insoluble metastannic acid is produced, negating the possibility of tin analysis. A mixture of fluoroboric acid, hydrogen peroxide, and disodium EDTA is used in the procedure to overcome these problems. Sample sizes range from 20 to 140 mg for lead alloys containing 40 and 2% tin, respectively. A hollow-cathode lamp and air–acetylene flame are used. The author prefers an electrodeless-discharge lamp and a nitrous oxide–acetylene flame.

Reagents and Equipment

A Perkin-Elmer Model 403 atomic absorption spectrometer and a Perkin-Elmer Intensitron tin hollow-cathode lamp were used for all samples. An air–acetylene flame and standard conditions were employed.

The solvent mixture was prepared by mixing equal volumes of reagent grade fluoroboric acid (48%), hydrogen peroxide (30%), and EDTA (0.2M).

Procedure

Weigh the samples in microsize (4 × 4 cm) disposable weighing trays. Use a balance that has a direct digital readability of 0.01 mg or, better, transfer the material from the trays, with the aid of water from a wash bottle, into 100 ml volumetric flasks and add 15 ml of the fresh solvent mixture. After the initial reaction subsides, heat the clear solutions on a steam bath for 5 min to decompose the excess hydrogen peroxide. Then cool the samples to room temperature and dilute to volume. Prepare the standard curve from a tin stock solution (reagent-grade tin metal in hydrochloric acid). Carry the final dilutions of these standards containing 15 ml solvent for every 100 ml of the final solution through the heating step prior to dilution. Use the instrument parameters recommended by the manufacturer. The standard curve is linear up to 150 $\mu g/ml$ tin.

Determination of Arsenic, Selenium, Tellurium, and Tin in Copper by Hydride Evolution (149)

Comment on the Method

Copper has been reported by most workers as a serious interferent in hydride generation work. In this method, copper is separated by precipitation of arsenic, selenium, tellurium, and tin using lanthanum hydroxide as a collector. A hydride reaction time of 20 sec is used to maximize collection of tellurium; the other elements do not require careful control of this parameter. Interferences are minimal. It is possible to determine 1 μg of each element in a hundredfold excess of each of the others without interference. Bismuth can also be analyzed by a similar procedure. Detection limits are (10 gm sample) arsenic 0.004, bismuth 0.002, tin 0.001, selenium 0.006, and tellurium 0.004 ($\mu g/ml$).

Reagents and Equipment

A Jarrell-Ash Model 82-526 spectrometer, an argon–hydrogen–entrained air flame, and a Perkin-Elmer arsenic–selenium sampling system to generate the hydrides were used. The gas flow settings were argon (15 liters/min), 25 psi; auxiliary argon (5 liters/min), 20 psi; and hydrogen (10 liters/min), 20 psi. Hollow-cathode lamps were used at 10 mA for each element, with the spectometer slit width 2 Å. Wavelengths were 1937 Å for arsenic, 2863 Å for tin, 1960 Å for selenium, and 2143 Å for tellurium. All solutions were prepared from analytical reagent-grade chemicals. Standard stock solutions (1 mg/ml) were prepared separately for arsenic, tin, selenium, and tellurium.

Arsenic stock solution.   Dissolve 1.3203 g of arsenious oxide in 2.4 M hydrochloric acid and dilute the solution to 1000 ml with 2.4 M hydrochloric acid.

Tin stock solution.   Dissolve 1.0000 g of tin metal as above.

Selenium stock solution.   Dissolve 1.4052 g of selenium dioxide as above.

Tellurium stock solution.   Dissolve 1.2507 g of tellurium dioxide as above.

Working standards.   These are made to contain 0.25, 0.50, 1.00, and 2.00 μg of arsenic, selenium, and tellurium separately, and 0.25, 0.50, 0.75, and 1.00 μg of tin per 20 ml of solution (0.5 M hydrochloric acid) were prepared from the stock solution. The appropriate volume of the stock solution was measured, 25 ml of 4 M hydrochloric acid were added, and the 200 ml volumetric flask was diluted to volume with distilled water. Lanthanum nitrate hexahydrate and sodium borohydride reagents were the commonly available grades.

Procedure

Transfer 10.0 g of copper, accurately weighed, to a 500 ml beaker. Add 25 ml of distilled water and dissolve the copper by adding slowly 45 ml of concentrated nitric acid. Add 10 ml of 5% lanthanum nitrate solution, dilute to 300 ml with distilled water, and add 100 ml of concentrated ammonium hydroxide. Let stand for 1 min, then filter the solution through a Whatman No. 41 filter paper. Wash the beaker and paper with 5% ammonium hydroxide solution and finally with distilled water.

Redissolve the residue into the original beaker with 40 ml of 6 N hydrochloric acid, washing the paper well with distilled water. Dilute the solution to 100 ml Reprecipitate by adding 30 ml of concentrated ammonium hydroxide. Let stand for 1 min, filter, and wash as before.

Dissolve the residue with 25 ml of hot hydrochloric acid (4 M) and wash with water. Transfer to a 200 ml volumetric flask and dilute to volume with distilled water. A reagent blank is processed through the whole procedure.

Transfer a 20 ml aliquot to the hydride generator flask. In the case of arsenic, selenium, tellurium, or bismuth determinations, add 7.5 M

hydrochloric acid: final acidity 4 M. For tin determination, add 20 ml of distilled water: final acidity 0.25 M. The final volume in all cases is 40 ml.

Connect the flask to the gas flow system with the hydrogen flame operating. Open the four-way stopcock for about 20 sec to admit argon, which flushes the air out of the system. After flushing, close the four-way stopcock and add a single sodium borohydride pellet via the dosing stopcock. Allow the reaction to continue for 30 sec, or 20 sec for tellurium and bismuth, using a stopwatch. Open the four-way stop-cock, which admits argon and sweeps the generated gases into the burner. Record the absorption signal on a recorder.

Standards, including a reagent blank, are measured using the same procedure to give calibration curves respectively for tin (0–1 μg) and arsenic, selenium, and tellurium (0–2 μg).

Determination of Silver, Arsenic, Bismuth, Antimony, Selenium, and Tellurium in Chromium Metal (150)

Comment on the Method

An oxidizing fusion with sodium peroxide is used for sample decomposition. Nickel is added to the sample in the furnace to prevent loss of analyte elements during charring. The charring temperatures recommended are much higher than those used by the author. The analyst may wish to try lower values. Analysis must be done using the method of standard additions.

Accuracy was tested by salting the sample with analyte prior to the fusion. Recoveries of 93–105% are recorded. A number of samples were also analyzed spectrographically, acceptable agreement being obtained with atomic absorption results except in the case of bismuth.

Reagents and Equipment

A Perkin-Elmer Model 403 atomic absorption spectrometer with a deuterium background corrector, the HGA 2000 graphite furnace, and Model 165 recorder were used for all analyses. All furnace temperature programs were measured using the calibrated current meter provided with the instrument. A hollow-cathode lamp was used for silver and electrodeless-discharge lamps served as light sources for arsenic, bismuth, antimony, selenium, and tellurium. A spectral slit width of 7 Å was used for all elements except bismuth, for which 2 Å was used. The graphite furnace tube was purged with argon; flow was interrupted automatically during atomization. Solutions were introduced to the graphite furnace by Eppendorf pipets and were dried at 125°C for 30 sec.

A stock solution of silver (1000 μg/ml) was prepared by dissolving high-purity metal in dilute nitric acid. Stock solutions of arsenic, bismuth, antimony, selenium, and tellurium (1000 μg/ml) were purchased from Harleco. Dilute standards were prepared from stock solutions as required. Nickel nitrate solution was prepared by dissolving high-purity nickel in dilute nitric acid. Analytical reagent-grade sodium carbonate and sodium peroxide were used for sample fusion. Distilled water was used for all dilutions.

Procedure

Prepare a blank. Transfer 1000 g of spec-pure chromium metal (60 M×D or finer) to a 50 ml zirconium crucible containing 6.0 g of sodium peroxide and 2.0 g of sodium carbonate. Mix thoroughly and cover with 1.0 g of sodium peroxide. Carefully fuse, first using a low flame and then increasing the heat until the fusion becomes molten and cherry red. Remove from the heat, cover, and allow to cool. Gently tap the crucible and transfer the fusion to a stainless steel beaker; cover and add 40–50 ml of water to the fusion and enough hot water to fill the crucible. Rinse contents of the crucible into the beaker and carefully boil the solution for 5 min. Cool and transfer to a 400 ml beaker. Acidify with 20 ml of nitric acid and boil for 10 min. Cool and transfer to a 200 ml volumetric flask, dilute to volume with water, and mix. Samples are prepared similarly.

Add a 20 ml aliquot of blank solution to a 25 ml volumetric flask. Transfer three 20 ml aliquots of the sample solution to three 25 ml volumetric flasks. Add 4 ml of nickel nitrate solution to each flask. Add standard-addition increments of element to the second, third, and fourth flask, and dilute all flasks to volume with water. Carefully adjust the instrumental conditions, bring the hollow-cathode lamp or electrodeless-discharge lamp beam position into best alignment with the deuterium arc beam. Inject 10 μl of the blank and sample solutions into the graphite furnace. Initiate the appropriate furnace program (Table XLVIII) for dry, char, and atomization cycles. Record the atomization response on the strip-chart recorder in terms of peak height for each 10 μl injection. Calculate the concentration of each element by the method of additions.

Table XLVIII

HGA 2000 Graphite Furnace Operating Conditions

Element	Wavelength (Å)	Char Temp °C Time (Sec)	Atomization Temp °C Time (Sec)
Ag	3281	1000° (30)	2400° (10)
As	1937	1500° (30)	1700° (10)
Bi	2231	1000° (30)	2300° (10)
Sb	2176	1500° (30)	2700° (10)
Se	1960	1500° (30)	2700° (10)
Te	2143	1500° (30)	2700° (10)

Analysis of Lead, Bismuth, Selenium, Tellurium, and Thallium in High-Temperature Alloys (151)

Comment on the Method

Furnace atomic absorption is used without prior separation of the analyte from the matrix. The char temperature for selenium is higher than that used by the author and the analyst may wish to experiment with lower values. Dissolution is accomplished in a mixture of hydrofluoric and nitric acids. The deuterium arc background corrector must be employed. The method was tested on standard alloys and good agreement obtained for the elements tested.

Reagents and Equipment

A Perkin-Elmer Model 403 atomic absorption spectrometer with deuterium background corrector, the HGA 2000 nonflame atomizer, and Model 056 recorder were utilized for all absorbance measurements. All temperatures reported here were measured using the calibrated current meter provided with the instrument. Standard hollow-cathode lamps were used for lead, bismuth, tellurium, and thallium, while a Perkin-Elmer electrodeless-discharge lamp served as the selenium light source. The atomizer tube was purged with argon. Solutions were introduced into the graphite furnace by Eppendorf automatic micropipets.

Stock solutions of lead, bismuth, tellurium selenium, and thallium were prepared by dissolving high-purity metals in dilute nitric acid. Solutions containing less than 0.1 μg/ml were prepared daily. Analytical reagent-grade acids were used for sample dissolution and stock solutions preparation, and deionized water was used for all dilutions.

Procedure

Dissolve 1 g of metal chips in 30 ml of a 1:1:1 mixture of nitric acid, hydrofluoric acid, and water by warming in a Teflon beaker. When dissolution is complete, reduce the volume to approximately 5 ml by evaporation. Cool, add about 20 ml of water, and heat to dissolve all salts.

Cool, transfer to a 50 ml plastic volumetric flask, and dilute to the mark with water.

Prepare standards by doping either well-characterized samples of base alloy or by combining proper proportions of stock solutions of the matrix elements to approximate the composition and concentration of the samples.

Select the instrumental conditions as indicated in Table XLIX. Carefully adjust the hollow-cathode or electrodeless-discharge lamp beam position to bring it into best coincidence with the deuterium arc beam. The temperature programmer is operated in the automatic mode such that the purge gas flow is interrupted 8 sec prior to atomization and is

Table XLIX

Operating Parameters and Detection Limits

Element	Pb	Bi	Te	Se	Tl
Wavelength (Å)	2833	2231	2143	1960	2769
Spectral band width (Å)	7	2	7	7	7
Solution volume (μl)	50	20	50	50	50
Sample concentration (mg/ml)	20	20	20	20	4
Drying temperature (°C)	150	150	150	600	150
Drying time (sec)	20	20	20	20	20
Char temperature (°C)	400	800	600	1000	500
Char time (sec)	60[a]	45	60[a]	45	60[a]
Atomization temperature (°C)	2000	2200	2200	1400	2000
Atomization time (sec)	5	5	5	5	5
Detection limit (μg/ml in sample)[b]	0.1	0.1	0.2	0.1	0.1

[a] Interrupt purge gas flow during initial 30 sec of char cycle.

[b] Detection limit is defined as the concentration of analyte in the sample that produces a peak signal-to-background ratio of 2.

not resumed until the atomization cycle is completed. Atomize samples and standards and prepare a graph of absorbance versus trace element concentration.

Determination of Traces of Lead, Bismuth, Thallium, Selenium, Tellurium, and Silver in Iron- and Nickel-Based Alloys (152)

Comment on the Method

Atomization is accomplished in a graphite furnace. Techniques involving separation of the trace elements from the matrix were compared with the direct-solution injection approach. The latter was chosen. Matrix effects were overcome by using the standard addition method.

Samples were dissolved by a variety of techniques. Dissolution in nitric acid proved satisfactory for low-alloy steels with hydrofluoric–nitric acid being best for nickel alloys. Lead contamination problems from reagent-grade (ACS) nitric acid were significant. The use of J. T. Baker Ultrex grade acid minimized this effect. Distilled water was passed through ion exchange and ultrafiltration beds to eliminate trace contaminants.

The nitric acid procedure was tested on NBS Series 360 low-alloy steels and iron. Good results were demonstrated for the certified elements, lead, bismuth, and silver. The hydrofluoric–nitric acid procedure was tested on NBS high-alloy nickel alloys. Satisfactory results were obtained for lead, bismuth, silver, titanium, selenium, and tellurium.

Reagents and Equipment

A Perkin-Elmer Model 503 atomic absorption spectrometer equipped with a Model HGA 2000 graphite furnace was used. The instrument was fitted with a smoke removal device, which consisted of two glass tubes mounted out of the light path at each end of the furnace housing and connected with flexible tubing through a Y joint to a water aspirator. In addition, an overhead exhaust system was used.

Simultaneous deuterium arc background correction was used. Careful alignment of hollow-cathode or electrodeless-discharge lamps with the background continuum source was accomplished with a center-zero strip-chart recorder, as described by the instrument manufacturer.

The analytical peaks were recorded on a strip-chart recorder capable of 10, 5, 4, 3, 2, and 1 mV full-scale response. Samples were introduced into the furnace by means of Eppendorf micropipets. Electrodeless-discharge lamps were used for all elements except silver, for which a hollow-cathode lamp was employed. The purge gas was argon.

Stock standard solutions were prepared from high-purity metals or oxides. In some cases commercially available standard solutions were employed. Working standard solutions of 10 μg/ml or less were prepared daily from stock standard solutions.

Nitric acid of the highest available purity (J. T. Baker Ultrex) was used for the analysis for lead of NBS 360 series iron and steel standards. Hydrofluoric acid was ACS certified grade. Water for the final dilution of samples was distilled, and then passed through three ion exchange resin beds and a microfiltration column. Other reagents were ACS certified.

Procedure

Nitric acid medium. Dissolve iron-base samples in dilute nitric acid; as much as 2 g of mild steel can be readily contained in a final dilution volume of 50 ml when 20% (v/v) nitric acid is employed.

Hydrofluoric acid–nitric acid medium. Dissolve high-temperature alloys in hydrofluoric acid–nitric acid: 20 ml of hydrofluoric acid, 10 ml of nitric acid, and 10 ml of water. This mixture readily dissolves a wider variety of high-alloy materials, including nickel–chromium base alloys.

Lead, bismuth, silver, and thallium can be determined at low and sub-ppm levels in mild steels and iron by direct introduction into the furnace of a dilute nitric acid solution of the sample. Dissolve 2.000 g in 20% (v/v) nitric acid and dilute to 50 ml.

Table L

Instrument Parameters[a]

Parameter	Pb	Bi	Tl[b]	Se	Te[b]	Ag
Wavelength (Å)	2833	2231	2768	1960	2143	3281
Bandwidth (Å)	7	2	7	2	2	7
Drying temp (°C)	150	150	150	150	150	100
Drying time (sec)	20	20	20	20	20	60
Charring temp (°C)	400	800	500	350	500	150
Charring time (sec)	60	60[c]	60[c]	60	60[c]	60
Atomizing temp (°C)	2000	2200	2000	2400	2000	2400
Atomizing time (sec)	8	8	8	10	8	8

[a] Aliquot volume: 5.0–20.0 μl.

[b] Maximum temperature for 30 sec after each measurement.

[c] Gas interrupt for first 30 sec of char cycle.

Once the linear range for each element has been established, by calibration curves prepared from pure iron and the element of interest, a spiking technique can be used. Add spikes directly to a second sample, which is treated identically. Use a spike that will yield an absorbance of about 0.5 above that of the sample alone. The absorbance must remain in the linear range. The final aliquot size is adjusted accordingly. At extremely low concentration levels, spikes are chosen to yield an absorbance at least 0.040 absorbance units above the absorbance of the sample.

Several important precautions must be taken. A precise knowledge of the linear range of each element in a given acid medium must be obtained so that spiked and unspiked samples are adjusted to fall in the linear absorbance range. A good estimate of the reagent blank must also be obtained.

The same approach is used for high-alloy materials, including nickel-based alloys, except that they are dissolved in hydrofluoric acid–nitric acid. The parameters used for these methods are given in Table L.

6
Analysis of Air Samples

COLLECTION AND TREATMENT OF
PARTICULATES IN AIR SAMPLES

Air is normally sampled for particulates using high-volume filters. Filter materials are usually cellulose or glass fiber. Air flow resistance is much higher in cellulose filters; hence a smaller volume of air will be sampled in a given period of time. Collection efficiencies of filters are generally unknown.

Air volume passing through filters is affected by pressure and temperature and by the fact that flow rate decreases with time. Corrections for these aberrations are generally not made.

The actual sample for analysis is cut as a disk from the large filter. Care must be taken to avoid contamination from the cutting tool. It is important to obtain a sample representative of the total filter surface. To this end it may be important to use results for several samples from one paper to obtain the final result.

The decomposition procedure used will depend on the composition of the filter, the nature of the particulate, and the requirement of the analysis. For most work with glass fiber filters, a decomposition involving hydrofluoric acid is recommended. In this way the major matrix constituent is volatilized, leaving a relatively simple sample solution.

Dry ashing of filters containing particulate material is suspect. There have been demonstrated losses of volatile elements such as lead, zinc, and cadmium even at temperatures as low as 500°C. The analyst must therefore ensure that ashing losses are negligible for his samples if dry ashing is employed. Wet ashing using a nitric–perchloric acid mixture can be used as an alternative.

Blank filters contain trace amounts of many elements. A distilled water rinse of glass fiber filters is generally employed to remove surface contaminants. It is common to find that a single batch of filters has similar contaminant composition. Filters from different batches usually contain widely varying levels of contaminating trace elements. Blank samples must always be included in a sample set.

Determination of Metals in Air Particulate Matter (153)

Comment on the Method

The procedure applies to the analysis of iron, copper, nickel, cadmium, lead, vanadium, chromium, and manganese in particulate material collected on glass fiber filters. Organic matter is removed at 500°C, the author recommends 400–450°C. Samples still should be tested for volatility losses. Dissolution is accomplished in a hydrofluoric–nitric mixture. (A low-temperature plasma arc asher was tried, but results were low by 60% for chromium and up to 9% for all other metals.) Matrix standards are used.

The procedure is designed for routine analysis of the nine metals. It is very important that sample decomposition is complete (visual observation for residue). Otherwise there is a risk of losing lead in the particulate.

The method is tested by the determination of the recovery of known amounts of metals added. The potential shortcomings of this method for testing the accuracy of a procedure must be borne in mind.

Reagents and Equipment

A stock matrix solution is prepared by dissolving 126.7 g Al $(NO_3)_3 \cdot 9H_2O$, 88.6 g $Ca(NO_3)_2 \cdot 4H_2O$, 4.75 g MgO, and 41.0 g $NaNO_3$ in 1000 ml of 3% nitric acid solution. The working standard solutions of copper, nickel, cadmium, chromium, vanadium, and lead are prepared in 1:10 dilution and iron and zinc in 1:100 dilution of the stock matrix solution.

A composite 1000 ppm stock solution of copper, nickel, cadmium, manganese, vanadium, and chromium is prepared by dissolving 1.000 g of each metal in a minimum quantity of concentrated nitric acid, except for chromium, which is dissolved in concentrated hydrochloric acid. The solutions are transferred to a 1000 ml volumetric flask and made to volume with demineralized, distilled water.

A composite 1000 ppm iron and zinc stock solution is similarly prepared by dissolving the metals in hydrochloric acid.

A 1000 ppm stock solution of lead is prepared separately by dissolving the high-purity metal in minimum quantity of nitric acid and making the volume to 1 liter with demineralized distilled water. A.R. grade soluble salts of metals, preferably nitrates, may also be used in place of high-purity metals.

The recommended concentrations (μg/ml) of working standards are as follows: Copper, nickel, cadmium, chromium, manganese, vanadium, lead: 0.0, 0.2, 0.5, 1.0, 2.0, 3.0, 5.0; zinc: 0.0, 0.5, 1.0, 2.0, 3.0, 5.0, 10.0; iron: 0.0, 20.0, 30.0, 40.0, 60.0, 80.0, 100.00; lead: 0.0, 7.0, 10.0, 15.0, 20.0, 40.0, 60.0, 80.0, 100.0.

A Perkin-Elmer Model 403 atomic absorption with air–acetylene and nitrous oxide–acetylene capability was used. An aluminum block 26 × 6 × 11 cm was drilled with 40 holes to accommodate 18 × 150 mm glass test tubes to a depth of 3 cm. Glass test tubes were marked at the 15 ml level. Then pure 100 ml Teflon dishes were used. Glass fiber filters 203 × 253 mm were employed (Mines Safety Appliance Company Ltd.).

Procedure

Cut a 19 × 253 mm sample strip and place on an asbestos board. Ash at 500°C in a muffle furnace for 2 hr. Transfer the strip to a 100 ml Teflon dish and add 15 ml of 1:1:1 hydrofluoric acid, nitric acid, and water. Take to a bone-dry stage by slow evaporation. Add 5 ml of concentrated nitric acid and take to dryness once again. Wet the contents with 1 ml of concentrated nitric acid, add 20 ml of water, simmer on the hot plate until the volume is reduced to about 10 ml, and then transfer quantitatively to the calibrated test tube. Place 40 prepared sample solutions in the aluminum block and heat for 2 hr on the hot plate. Bring the volume up to the 15 ml mark at room temperature with demineralized, distilled water. Seal the test tubes with parafilm and mix. Make a 1:10 dilution for iron and zinc and analyze using matrix simulated standard solutions to calibrate the instrument.

Samples are run on the atomic absorption unit using the manufacturers recommended settings. Vanadium is done using a nitrous oxide–acetylene flame. The others are analyzed with an air–acetylene flame.

Calculation

$$\mu\text{g (metal)/m}^3 = \frac{(\mu\text{g/ml sample} - \mu\text{g/ml blank}) \times 140}{\text{air volume in m}^3}$$

Determination of Organic and Total Lead in the Atmosphere (154)

Comments on the Method

The authors define as organic lead that lead containing material that passes through an 0.45-μm porosity glass fiber filter. No proof is given for this statement in the original paper. Particulate lead is collected on the filter and organic lead passing through is collected in an iodine monochloride solution.

An aqua regia dissolution is used for the analysis of particulate lead. It must be emphasized that this will not necessarily yield total lead in particulate.

Lead trapped in the receiver solutions is extracted using ammonium pyrolidine dithocarbonate (APDC) into MIBK. It is important to use blank solutions with this procedure.

Reagents and Equipment

Iodine monochloride (1.0 M). To 800 ml of 25% (w/v) potassium iodide solution add 800 ml of concentrated hydrochloric acid and cool the mixture to room temperature. While stirring the solution vigorously, slowly add 135 g of potassium iodate. Continue the stirring until all free iodine has dissolved to give a clear orange–red solution. Cool the mixture to room temperature and dilute to 1800 ml.

Iodine monochloride (0.1 M). Dilute 100 ml 1.0 M iodine monochloride to 1 liter. This reagent is stable for an indefinite period under ambient laboratory conditions.

Hydrochloric acid–nitric acid mixture. Add one volume of distilled constant boiling (about 19%) hydrochloric acid to four parts 40% distilled nitric acid.

Buffer solution (3.4 pH). Dissolve 17 g of potassium acid phthalate in water. Add a 1.4 ml volume of concentrated hydrochloric acid and dilute the mixture to 1 liter with distilled water.

Standard inorganic lead solutions (1000 μg Pb/ml). Dissolve a 1.598 g weight of dried lead nitrate in 1 liter of 0.5% nitric acid. Prepare working solutions of 1, 5, and 10 μg Pb/ml daily by appropriate dilution of stock solution with 0.5% nitric acid.

Standard organic lead solution (1000 μg Pb/ml). Dissolve a 0.2437 g weight of lead, N,N-diethyl dithiocarbamate in 100 ml of methyl isobutyl ketone (MIBK) and store in brown reagent bottle. Prepare working solutions of 0.5 and 1 μg Pb/ml daily by appropriate dilution of stock solution with MIBK. Prepare a fresh stock solution weekly.

Gas samples. A sampling device developed by the Environmental Protection Agency and manufactured by the Research Appliance Corp.

(EAC, Alison Park, Pennsylvania) was used for collecting air samples. The gas sampler consists of a glass inlet manifold, membrane prefilters, tubes containing appropriate collecting solutions through which a sample can be passed, an air trap, membrane exhaust filters (to protect critical orifices), an outlet manifold, and an external vacuum pump. Number 20 gauge hypodermic needles were used as critical orifices in order to obtain a 2.5 m^3 air sample at a constant flow rate of about 2 liters/min for a 24 hr period. The traps for the gas sampler were filled with activated carbon (10–20 mesh) in order to protect the critical orifices from iodine monochloride vapors. Glass tubing with a 1 to 2 mm i.d. opening was used as dispersers for the air stream. This sampling device, currently in use by the Environmental Protection Agency, allows collection of up to five samples simultaneously. Any sampling train consisting of a prefilter, an air scrubber, a trap, an air-measuring device, and a vacuum pump is suitable for this method.

Membrane filters.   Millipore type HA (0.45 μm pore size) were used.

Separatory funnels.   These were 125 ml capacity with Teflon stopcocks and stoppers.

Procedure

Add 50 ml of 0.1 M iodine monochloride to each of the sample tubes in the gas sampler. Connect a membrane filter to the inlet of each tube and connect the vacuum pump to the outlet manifold of the gas sampler. Determine the flow rate through each tube with a wet-test meter. Sampling is done for 24 hr, resulting in a collection of approximately 2.5 m^3 of air. Redetermine the flow rates with the wet-test meter.

Particulate lead analysis.   Remove the membrane filter with care from the holder and place in a 100 ml beaker. Add 5 ml of water and 5 ml of the hydrochloric–nitric acid mixture and digest on a hot plate for 1 hr. Cool the solution. Dilute the mixture to 10 ml. Determine the lead concentration of each solution directly by atomic absorption spectrometry by comparing the absorbance of each sample extract with the absorbance of standard lead solutions made up in 0.5% nitric acid.

The number of micrograms of lead per milliliter of sample extract is multiplied by the appropriate dilution factor and divided by the number of cubic meters of air represented by the sample. The cubic meters of air sampled from the average of the two flow measurements taken before and after sampling are determined.

Organic lead analysis. Disconnect each of the tubes containing the iodine monochloride solution from the gas sampler. Empty each sampling solution into a 150 ml beaker and rinse the disperser, tube, and cap with distilled water into each beaker. Prepare two blank iodine monochloride solutions by pouring 50 ml portions of 0.1 M iodine monochloride into each of two 150 ml beakers. Make two 5 μg Pb/ml standards by adding 1 ml of 5 μg/ml standard to each of two 50 ml portions of iodine monochloride solution. Titrate all blanks, standards, and samples with 20% sodium sulfite to a colorless endpoint. Add two drops of methyl orange indicator. Titrate the solution with 10% ammonium hydroxide to a yellow endpoint. Add 10 ml of 3.4 pH buffer to each solution followed by 2 ml of 1% APDC and 1 ml of 1% hydroxylamine hydrochloride. Measure the pH of each solution and adjust to 3.6 $\pm$ 0.1 with dilute hydrochloric acid or dilute ammonium hydroxide if necessary. Transfer each sample to a 125 ml separatory funnel, rinsing each beaker with distilled water. Add 5 ml of MIBK. Shake each separatory funnel for 1 min and drain the aqueous layer off and discard. Pack the stem of each separatory funnel loosely with a small piece of cotton and drain the remaining MIBK into a small container that can be tightly capped.

Analyze each of the MIBK solutions on the atomic absorption spectrometer according to the procedure recommended by the instrument manufacturer for analysis of organic solutions. Adjust the nebulizer for optimum response using a 1 μg/ml organic lead standard in MIBK. Determine the amount of lead in the MIBK extracts of each sample by comparing the absorbance of each sample extract with the absorbance of MIBK extract of the lead standards in iodine monochloride prepared by the same procedure used with the samples. Both samples and standards are corrected for the average blank absorbance. Concentrations of organic lead are calculated dividing the number of micrograms of lead found in the MIBK extract by the volume of the sample in cubic meters.

Detection limit.   Organic lead, 0.2 μg/m^3; particulate lead, 0.4 μg/m^3.

Automated Procedure for Arsenic in Air Particulate (155)

Comments on the Method

This is an automated hydride method that could likely be applied to samples other than air particulate after an appropriate decomposition. Again, it would be important with other samples to make certain that the autoanalyzer hydride reaction times were optimized.

An extensive list of interferences was investigated: iron, aluminum, copper, cobalt, magnesium, nickel, zinc, cadmium, chromium, lead, manganese, vanadium, sodium, potassium, and selenium in maximum amounts likely to occur in air particulate (20 ppm). Of these, only selenium, cobalt, nickel, and copper caused serious interference. At the levels usually encountered in solutions of air particulate, <5 μg/ml, little interference will occur. Nonspecific absorption effects were investigated and found to be of no consequence.

The method was tested by spiking sample solutions. Good recoveries were obtained.

Reagents and Equipment

The following reagents were used: sodium borohydride (Fisher Scientific Co., 90% pure powder) as a 1% w/v solution in dilute sodium hydroxide; sulfuric acid (A.R. grade); nitric acid (A.R. grade); nitrogen gas; and 1000 μg/ml arsenic and atomic absorption stock standard (Fisher Scientific Co.). Working standards were prepared by serial dilution of the stock standards with a 1:10 dilution of a 2 + 1 sulfuric–nitric acid mixture.

A Varian-Techtron AA5 atomic absorption spectrometer with a recorder was used. Arsenic was measured at 1937 Å in the absorbance mode. Scale expansion of 2× was used on the recorder and the chart speed was 5 mm/min. The quartz cell temperature was 800 ± 20°C.

The sampler and proportioning pump used were Technicon modules available in the laboratory. The quartz tube furnace was a 10 cm long by 6 mm diameter tube with a 17.5 cm long and 4 mm diameter inlet tube fused in the center. The quartz cell was wound with approximately 2 m of 22 gauge asbestos-covered nichrome heating wire and further insulated by wrapping with asbestos tape and a layer of asbestos cord. Insulated terminals were provided for the service cord and plug. The finished cell was mounted on the burner assembly and secured with metal wire. This allowed the use of the burner alignment controls to position the cell on the optical path. The temperature of the heated cell was controlled by a Variac transformer. The complete combined manifold is illustrated diagrammatically in Fig. 40. The instrumental parameters are presented in Table LI.

Procedure

Collect air particulates on an 8 × 10-in. fiberglass filter. Punch a 0.4-in.² circle out of each test filter. Cut the punched sample into small pieces and place in an 18 × 150-mm borosilicate glass test tube with a calibration mark at 15 ml. Place 40 test tubes in a 10.5 × 9.5 × 1.5-in.

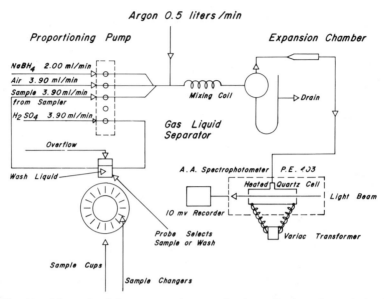

Fig. 40. Schematic of the automated system for determination of arsenic (155).

aluminum block with 40 holes. The block is used as a test tube holder as well as a heating device. Add 1.5 ml of a mixture of sulfuric and nitric acids (2:1 v/v) to each test tube. Heat the aluminum block containing the test tubes for 1 hr on a hot plate set at 250°C.

Add demineralized, distilled water to within 1 ml of the 15 ml mark on the test tubes. Cool the contents of the tubes to room temperature and bring the volume up to the 15 ml mark. Shake the test tubes vigorously to break up the filter, seal with parafilm, and centrifuge. Pour the solutions into sample cups and position in the sampler. Analyze

Table LI

Proportioning Pump and Automatic Sampler Parameters

Air	2.90 cm³/min
Sodium borohydride	2.5 cm³/min
Sample or wash	2.5 cm³/min
Sulfuric acid, 2.4 N	2.0 cm³/min
Mixing coil	29 turns, 1.8 cm diam.
Expansion chamber	10 cm³ cylindrical 5/8 × 3 in.
Cam auto sampler	2 lobe
Sample time	60 sec
Wash time	120 sec
Argon flow	270 cm³/min

blank filters similarly. In the case of samples low in arsenic, more than one circle is digested with acid.

Optimize the instrumental adjustment. Align the instrument and heated cell and allow to equilibrate, with the argon flowing and reagents pumping. The two-lobe cam on the sampler must be specially cut to allow enough time for the recorder pen to reach the baseline between sample pickups. Set the recorder pen to the baseline and switch the sampler on for unattended operation. Background correction is not needed.

Determination of Beryllium in Airborne Particulate (156)

Comments on the Method

Flame or flameless atomization may be used, with the latter giving a hundredfold better limit of detection (0.00005 μg/m^3 based on a 2000 m^3 sample). Collection is on cellulose or glass fiber filters, detection limit being the same with either matrix. No pretreatment of the filter is required with cellulose, whereas glass filters must be washed with repetitive treatments of distilled water.

Significant changes in the beryllium blank between different batches of filters were noted. For this reason, it is recommended that the blank be checked on a number of filters in each batch.

Reagents and Equipment

Glass filters, commercially available material in size 203 × 254 mm, are exhaustively washed prior to use. Cellulose filters, ashless, acid-washed, analytical grade, in size 203 × 254 mm are used. Water should be distilled at least twice from glass or quartz. Hydrofluoric acid, 49%, reagent grade, in polyethylene containers is employed. Nitric acid is 71% reagent grade.

Standard solution of beryllium.   Dissolve 22.757 g of beryllium nitrate tetrahydrate in water and make up to 1 liter. In 1 ml of this stock solution there is 1000 μg of beryllium.

An atomic absorption spectrometer, with meter, recorder, or digital readout, and monochromator with wavelength dial reading to 1 Å are employed. Use the 2348 Å beryllium line. Polyethylene bottles, screw cap, for storage of test samples of 30 ml capacity are used. A Büchner funnel, polypropylene is made with a sintered false bottom and vacuum connection, 216 × 267 × 85 mm in free depth. A Massman design graphite furnace should be used with a strip-chart recorder. Automatic Eppendorf pipets with capacities 10–100 μl are used.

Cellulose filters may be used without further treatment. Glass fiber filters are purified by placing a group of 100 such filters in the special Büchner funnel and extracting repetitively with distilled water.

Mount the filter in a conventional high volume or other sampler head. Draw air through the filter at a flow rate between 1.13 and 1.60 m^3/min, for an appropriate period, such as 24 hr. The resistance to flow offered by the cellulose filters is much greater than for glass, and an appreciably smaller total volume of air will be taken with cellulose. Calculate and record the total volume of air sample (in m^3) as the product of mean flow rate and time.

Procedure

Cut aliquots from the exposed surface of a filter using a circular metal punch. The cutting edge of the punch is carefully wiped with lens tissue between each use to prevent carryover of contamination from one sample to another. Place one or more such disks in a Teflon beaker. Initiate the dropwise addition of 1 ml of hydrofluoric acid. Gently warm the contents of the beaker at low heat until the hydrofluoric acid is almost completely evaporated. At this point, add 1–2 ml of nitric acid, and continue to heat gently until a few drops of nitric acid are left. Add about 10 ml of water, bring nearly to the boil, and filter through a Whatman No. 41 filter into a glass beaker. Transfer to a 25 ml volumetric flask. Rinse down the Teflon beaker with another 10 ml of water, warm, and filter into the same beaker. Transfer to a 25 ml volumetric flask and make up test sample to mark. Mix the contents of the volumetric flask thoroughly after adjustment to volume. Transfer contents of volumetric flask to a polyethylene storage bottle. The test sample is now ready for analysis. The above procedure applies to the digestion of glass fiber filters. Cellulose filters may be digested or extracted with nitric acid.

Flame (nitrous oxide–acetylene) analysis. Aspirate calibrating solutions over the concentration range 0–4 μg/ml with a blank solution between each standard. Likewise, aspirate sample solutions. Each solution should be run for 30 sec.

Furnace analysis. By means of an automatic pipet with polyethylene tip, place identical microliter volumes of the dilute standard beryllium solutions and a distilled water blank in the furnace. Measure and record the response of each test portion following the predetermined measuring cycle. Cover the range 0.2–10 ng of beryllium in steps of 0.2 ng

(0.01–0.05 µg/ml for a 20 µl sample). The recommended furnace parameters are drying, 20 sec, 100°C; charring, 20 sec, 1100°C; atomizing, 20 sec, 2400°C. Samples are run in the same manner.

Blanks of both reagents and reagents plus filters are run.

Determination of Cadmium, Copper, Manganese and Lead in Air Particulate by Direct Furnace Atomization (157)

Comment on the Method

A direct furnace atomization method for air filters was adapted from an existing method for the analysis of pulp and paper. The method is suitable for cadmium, copper, manganese, and lead.

The furnace used was an induction-heated home-made unit (Fig. 41). Its construction details are included in the method. Rapid atomization characteristics of lead and cadmium allow these elements to be determined by measuring peak heights. Integration must be used for copper and manganese. A standard addition technique was used for standardization.

The method was checked against a wet leaching, conventional, electrothermal atomization procedure and good agreement obtained. The latter method had been previously verified using NBS SRM 2676 metals on filter media.

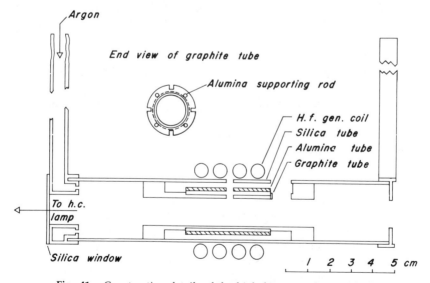

Fig. 41.   Construction details of the high-frequency furnace (157).

Reagents and Equipment

A Perkin-Elmer Model 303 atomic absorption spectrometer was used with a deuterium arc background corrector and two-channel recorder. One channel recorded percentage absorption and the other transformed the signals to absorbances and gave integrated peak areas.

The high-frequency induction-heated graphite furnace shown in Fig. 41 was used. The materials employed were 18.5 mm diameter graphite rods, quality RWO (Ringsdorff-Werke, Germany); alumina tubes quality A1 23, 20 mm o.d., 19 mm i.d., length 40 mm (Degussa, Germany); silica tubes, 28–29 mm o.d., 25–26 mm i.d. (Thermal Syndicate, Great Britain).

The furnace was attached to and heated by a Philips 2 kW high-frequency induction generator model MA 20, the temperature being adjusted with a variable steppless transformer (0–260 V). The maximum temperature was about 2000°C.

The generator box was placed on an adjustable and movable table. Weighings were made with semimicro or microbalances.

Atmospheric particulates were collected on 37 mm diameter Whatman No. 40 ash-free cellulose filter paper. The diameter of the exposed filter area was 25 mm. The air intake was connected via a manifold to eight filter holders made from high-density polyethylene. The filter holders were designed to give a homogeneous deposition of particles over the exposed filter area. Each filter holder was connected to a simple membrane pump (aquarium pump), and calibrated with a rotameter at the start and end of the sampling period. The flow rate of the pumps was about 150 liters/hr.

The nitric acid was reagent grade (Merck). The graphite furnaces were purged with argon of purity 99.9% (by volume). Deionized distilled water was used.

Primary standard solutions (1000 μg/ml) were prepared by dissolving high-purity metals in an excess of nitric acid and diluting to 1 liter. Highly diluted secondary standard solutions were prepared daily; nitric acid was added to maintain pH $\leq$ 2.

Procedure

Cut 2.5 and 5.5 mm disks from the filters with punch pliers.

Before a new furnace is put into use, record the transformer voltage/furnace temperature relationship with a thermocouple ($\leq$ 1500°C) and with an optical pyrometer (1500–2000°C). The temperature should be measured at a point on the interior of the bottom. Also register the time required to obtain thermal equilibrium.

Table LII

Operating Conditions

Wavelength (Å)	High-frequency furnace (sec/°C)		
	Dry	Ash	Atomize
Cd 2288	60/80	60/300	15/1600
Cu 3274	60/80	60/300	15/1900
Mn 2795	60/80	60/300	15/1900
Pb 2833	60/80	60/300	15/1600

Use an argon flow rate of 6 ml/sec. Introduce the sample into the furnace. Move the furnace into position. Run the samples using the parameters given in Table LII.

Add standard solutions to the filter disks using 1-μl pipets. Draw the standard addition curves by averaging the values from duplicate measurements of filters without and with two additions of standard solution. Correct the analytical signals for the blank signal given by the metal content in the filters. To reduce the blank level, the filters may be leached with dilute nitric acid and washed with distilled water before use.

7
Analysis of Petroleum and Petroleum Products

Trace metals in crude oils can poison the catalyst during the cracking of petroleum crude oils. This has the effect of decreasing catalyst efficiency and increasing the yield of unwanted by-products. It is important, therefore, to have sensitive analytical methods for the routine determination of trace metal content of crudes.

The metal content of lubricating oils increases during operation of a motor due to wearing of metal parts. Different alloys are used for the various components of an engine. Hence, by monitoring selected trace elements in lubricating oils it is possible to detect abnormal wear or to predict when a particular component should be routinely replaced in an engine. This is particularly useful in the aircraft industry, where air safety depends on, among other things, the ability to diagnose faulty parts prior to their failure when in use.

Metals occur in oils in a variety of ways. From the point of view of sampling it is important to realize that metals may occur in particulate material. This is generally the case with wear metals.

Sample preparation usually involves dilution of the viscose oil with a light solvent. Xylene is commonly used for this purpose. In addition, when particulate is present an acid addition may be necessary to solubilize the metals completely. Standards are normally prepared from organometallic compounds solubilized in the solvent.

Determination of Wear Metals in Engine Oils by Atomic Absorption Spectrometry with a Graphite Rod Atomizer (158)

Comment on the Method

The following metals may be determined by the proposed procedure: silver, chromium, copper, iron, nickel, lead, and tin. A comparison was made by the authors between their results and those obtained by other laboratories participating in the U.S. Air Force Spectrochemical Oil Analysis Program (SOAP). Good agreement was obtained for silver, copper, iron, nickel, and lead. The chromium results by the carbon rod technique were higher than those obtained by the air–acetylene flame method, but compare favorably with the nitrous oxide–acetylene flame procedure. Tin levels in the oils were found to be near or below the detection limit for flame atomic absorption and sometimes below the detection level with the carbon rod. Comparisons for tin were thus difficult.

It is important to emphasize that this method has been tested in an interlaboratory comparison. While only atomic absorption was used in the comparison and hence a systematic bias might be present, the good agreements obtained should give a worker more confidence in this procedure than is normally possible in the usual case where no accuracy assessment is made.

Reagents and Equipment

The graphite rod atomizer described by Molnar et al. (159) was used without further modification. The atomizer was mounted in place of the burner in a Perkin-Elmer 303 atomic absorption spectrometer equipped with a recorder readout unit and a Sargent TR recorder. The photomultiplier output was also connected to a digital integrator (Autolab 6230, Vidar Corp., Mountain View, California) with printout facility. Both the recorder and the integrator were operated in modes giving the most rapid response. During the course of this work, the use of an integrated absorption signal generally provided no increase of precision over that obtained by using peak absorbances taken from the recorder chart, and the use of the integrator was eventually discontinued.

Single- and multielement hollow-cathode lamps (Perkin-Elmer, Varian Techtron, Aztec) were operated at the manufacturers' recommended maximum currents.

Graphite rods were machined from Poco FX91 graphite (Poco Graphite Inc., Decatur, Texas), and a cylindrical cavity 1.4 mm in diameter and 1.0 mm deep was drilled in the center of the top of each rod. The cavity volume (about 1.5 μl) was suitable for containing and ashing the 0.3–0.8 μl samples used in this work.

Samples were dispensed from a Hamilton 1.0 μl syringe (7101 H-CH, Hamilton Co., Whittier, California) with a Chaney adaptor and a tungsten plunger extending to the tip of the needle. The use of this type of syringe minimized syringe cleaning problems that were occasionally encountered in withdrawing oil samples with syringes having a plunger extending only to the end of the barrel.

Power was supplied to the graphite rod from a 250 A, 10 V DC supply (SCR Power Supply, Electronic Measurements, Inc., Oceanport, New Jersey), with the current-time sequence controlled from a preset program in an adjustable timing circuit. The atomizer was used in conjunction with a flow of argon (7.2 liters/min) and hydrogen (1.2 liters/min). The argon–hydrogen–entrained-air flame that ignites when the rod reaches about 600°C is valuable in preserving the atom populations for a considerable distance above the rod.

The standards used in this work were solutions of organometallic compounds in a neutral-base oil (Phillips Petroleum Company Condor 105) supplied by SOAP. The following certified NBS organometallic compounds were used: silver 2-ethylhexanoate, tris(1-phenyl-1,3-butanedionato) chromium(III), bis(1-phenyl-1,3-butanedionato) copper(II), tris(1-phenyl-1,3-butanedionato) iron(III), nickel cyclohexanebutyrate, lead cyclohexanebutyrate, dibutytin bis(2-ethylhexanoate). About 10 solutions were used, containing each metal in the following concentrations: 0, 1, 2, 3, 5, 10, 15, 20, 50, and 100 μg/ml. The 50 μg/ml solutions contained only 20 μg/ml of silver while the 100 μg/ml solution contained no silver. Magnesium and aluminum were also present in all solutions. Solutions with 35 and 75 μg/ml of all elements (except silver) were also prepared from those provided.

Procedure

Place a suitable volume (0.3–0.8 μl) of a standard solution in the cavity on the graphite rod, and ash for 18 sec at about 440°C. Increase the current rapidly to a higher value for 3 sec to give a temperature high enough to atomize the element being determined. Check the absorption time record to ensure that the absorption returned to zero after completion of the ashing and before carrying out the atomizing step. Run all samples in duplicate, and in general, if the duplicates do not agree to within 4%, make a third determination.

Used jet engine oils are treated in an identical manner to the standards. The heavier reciprocating-engine oils require an additional 5 sec of ashing at about 500°C to remove the less volatile organic constituents.

Table LIII gives details of instrumental conditions, sample sizes, and optimum concentration ranges for each element. It should be noted that it is not always necessary or desirable to operate under conditions giving the maximum sensitivity.

Several different methods of reducing the sensitivity can be used when necessary. The methods available include dilution of the standards and samples with an organic solvent such as isooctane or methylisobutyl ketone; use of a less-sensitive spectral line; use of a

Table LIII

Experiment Conditions

Element	Wavelength (Å)	Atomizing temperature (°C)[a]	Sample volume (μl)	Concentration range (μg/ml)	Sensitivity reduction[b]
Ag	3281	1400	0.5	0–5	(iv)
Cr	3579	1920	0.5	0–10	—
Cu	3247	1640	0.5	0–30	(i), (iv)
Fe	3720	1920	0.5	0–50	(i), (ii), (iv)
Ni	2320	2050	0.5, 0.8	0–5	—
Pb	2833	1440	0.5, 0.3	0–100	(iii), (iv)
Sn	2863	1600	0.8	0–5	—

[a] All standards and type-A jet engine oils were ashed for 18 sec at 440°C. All type-B jet engine oils were ashed for 23 sec at 440°C. A drying time of 10 sec was used.

[b] (i) Solutions diluted with isooctane, (ii) less-sensitive spectral line used, (iii) smaller sample volume, (iv) graphite rod lowered.

smaller sample volume; and increasing the vertical distance between the graphite rod and the light beam from the hollow-cathode lamp. Each method has certain disadvantages or limitations. Dilution is time consuming, and vitiates one of the main reasons for using flameless atomization in the first place; suitable less-sensitive lines are not available for all elements; the use of sample volumes smaller than about 0.3 μl increase the relative error of volume measurement and dispensation and increase the range of nonrepresentative sampling; lowering the graphite rod until it is 3–7 mm below the bottom of the beam from the hollow-cathode lamp is satisfactory, but beyond this region reproducibility may be severely impaired. By using a Hamilton 1.0 μl syringe with Chaney adaptor, syringe cleaning is simple and no air or vacuum bubble problems arise. Also, the reproducibility of measurements on 0.5 μl oil samples is as good as that for the standards, i.e., the relative standard deviation was 4% or less.

The atomizing temperatures quoted in Table LIII are those reached in the cavity at the end of the 3 sec atomization step and are therefore higher than the temperatures at which atomization actually occurs.

Determination of Titanium in Lubricating Oils (160)

Comment on the Method

A detection limit of 0.03 ppm is reported for this method using the nitrous oxide–acetylene flame. The oil is diluted and shaken with methylisobutyl ketone containing 0.15 ml of a 1:3 v/v mixture of hydrofluoric–hydrochloric acid. This acid mixture is necessary for the complete dissolution of titanium particles in the oil. The method is rapid, 1–2 min total analysis time, and has given good results when used on U.S. Air Force standard lubricating oils.

Reagents and Equipment

A Perkin-Elmer 305B atomic absorption spectrometer equipped with a corrosion-resistant nebulizer was used. The output signal was fed into a Leeds and Northrup Speedomax W strip-chart recorder. In all cases a scale expansion of 30× was used for concentrations less than 10 ppm and 3× was used for 10–100 ppm with an integration time of 10 sec. Ten chart divisions are equivalent to one absorbance unit; therefore, absorbance was obtained from chart divisions/(10× scale expansion). The fuel/oxidant flowmeter setting was 5.0/6.0, which is equivalent to (6.5 liters/min)/(12.5 liters/min). The flowmeter settings were obtained directly from the flowmeters of the burner control box, a Perkin-Elmer Model 303-0678. However, they were converted into flow rates in liters per minute using the manufacturer's reported plot of flowmeter reading vs. liters per minute. Samples were contained in 1 oz polyethylene bottles fitted with polyethylene screw cap lids. A plastic pipet and Lab Industries Repipet Jr. were used to dispense the required amounts of acid solution.

The acid solution required was prepared by mixing by volume one part reagent-grade concentrated (48%) hydrofluoric acid and three parts reagent-grade concentrated (37%) hydrochloric acid. Mobil (mil spec 7808) unused aircraft lubricating oil was used as the blank and also to dilute the Conostan 5000 μg/ml concentrate in order to prepare a series of standards covering the 1–100 ppm range. A 325 mesh titanium powder obtained from Metal Hydrides Company (now Ventron Corporation) was used to prepare other standards used in this work. A quantity of 43.14 mg titanium powder, suspended in blank oil, was used to prepare 265.7 g of a 200 ppm stock solution that was contained in a polyethylene bottle. Then 1 ml of the acid solution was added to the stock solution and the mixture was shaken vigorously for 5 min to quantitatively dissolve the titanium powder. Aliquots were taken to prepare a series of standards in the 1–100 ppm range. The samples studied in this work were used aircraft lubricating oils, obtained from Dover, McConnell, and Kelly Air Force Bases, which were also analyzed by AES in the SOAP laboratories and found to contain titanium. The Conostan titanium standard is an alkyl/aryl sulfonate of titanium obtained as a concentrate of 5000 ppm from the Conostan Division, Continental Oil Co., Ponca City, Oklahoma. The Special D12 standard is a mixture of Conostan silver, aluminum, chromium, copper, iron, magnesium, molybdenum, nickel, lead, silicon (not sulfonate), tin, and titanium alkyl/aryl sulfonates prepared in lightweight hydrocarbon oil by the Naval Air Rework Facility, Naval Air Station, Pensacola, Florida.

Procedure

Dilute 5 g of a used turbine engine oil sample with 10 g MIBK and 0.15 ml of 1:3 by volume, hydrofluoric–hydrochloric acid solution in a 1 oz. polyethylene bottle. Shake the mixture vigorously by hand for 10 sec. The blank oil and the standards are prepared in the same manner, and the analysis is accomplished according to the conditions listed in Table LIV. The concentrations in ppm for the oil samples are obtained from the working curve.

Determination of Vanadium and Nickel in Oils Using Electrothermal Atomization (161)

Comment on the Method

The oil samples are diluted using xylene. The use of sample volumes greater than 10 μl results in poor reproducibility. If larger volumes must be used, a number of 10 μl portions are injected with a drying cycle following each injection.

Background correction must be employed for nickel analysis. This technique was not necessary in dealing with vanadium. The paper does not record any other interference studies.

Table LIV

Instrumental Parameters

Wavelength	3656 Å
Slit	2 Å
Grating	uv
Titanium lamp	40 mA
Nitrous oxide	
Tank regulator	40 psig
Burner control	30 psig
Flowmeter setting	6.0 (center of ball)
Acetylene	
Tank regulator	12 psig
Burner control	8 psig
Flowmeter setting	5.0 (center of steel ball)

Reagents and Equipment

A Perkin-Elmer Model 303 atomic absorption spectrometer equipped with a Perkin-Elmer (Bodenseewerk) graphite cell HGA 70, an automatic recorder readout accessory, and a Hitachi–Perkin-Elmer Model 150 recorder was used. A deuterium arc background corrector was applied in the procedure for nickel. Small sample aliquots were injected into the graphite tube by means of Eppendorf-Marburg micropipets. Flame atomic absorption measurements were carried out on the instrument equipped with a Boling burner head. Vanadium and nickel single-element hollow-cathode lamps were used as light sources. The xylene was laboratory reagent quality (Merck) and other chemicals were reagent-grade quality.

Stock solutions of nickel and vanadium (100 μg/ml) were prepared from ammonium vanadate and nickel sulfate hexahydrate dissolved in 0.1 M nitric acid. More dilute (1.0 and 0.1 μg/ml) standard solutions containing 5 ml of nitric acid per 100 ml were prepared daily by suitable dilution.

Before samples were atomized in the graphite tube, any solvent, organic material, or other unwanted component of the matrix was removed by heating the tube in stages to an appropriate temperature; otherwise false signals could be obtained. The optimal times and temperatures of the various stages were found to be as follows: the sample solutions were dried at 100° and then charred at 1100°C by means of program 7 on the HGA 70 with a 30 sec sequence duration; for the atomization stage 2400° (8 V) was selected for nickel and 2700° (10 V) for vanadium.

Procedure

Weigh out a suitable quantity, e.g., 1 g, of oil into a 10 ml volumetric flask and dilute to volume with xylene. In the case of pollutants contaminated with insoluble material, centrifuge or filter the suspension, take 2 ml in a platinum crucible, and evaporate the xylene on a hot plate at very low heat; weigh the oil left in order to evaluate the quantity dissolved.

For vanadium inject 5–50 μl of 1 μg/ml standard solution into the graphite tube and measure the absorption peak using program 7 (10 V) atomization voltage, 30 sec sequence duration, and the 3182 A analytical line.

For nickel inject 5–50 μl of 0.1 μg/ml standard solution and record the peak using program 7 (8 V) atomization voltage, 30 sec duration, and the 2320 Å analytical line.

Record blank values at the same instrumental settings and subtract them from the analytical value after conversion to absorbance. Either 20 ng of vanadium or 5 ng of nickel produce about 20% absorption at scale expansion 1. Prepare calibration graphs and measure 10 μl sample volumes in the same way. If necessary, inject several 10 μl aliquots successively before the charring and atomization steps but always dry the solvent at 100° between each loading.

For vanadium calculate the concentration directly from the calibration graph after converting peak heights to absorbance and subtracting the blank. For nickel, make a new absorption peak reading at 2316 Å and subtract the scattering signal from the analytical value before calculating the concentration by means of the calibration curve. If available, use a deuterium arc background corrector to avoid measuring at two different wavelengths.

Determination of Lead in Petroleum and Petroleum Products Using Electrothermal Atomization (162)

Comment on the Method

The sample is diluted with xylene and then the lead extracted, using dithizone, into 40% HNO_3. No interference was detected from thousandfold excesses of Ba^{2+}, Ca^{2+}, Ce^{3+}, Co^{2+}, Cr^{2+}, Fe^{3+}, Mg^{2+}, Na^{2+}, Ni^{2+}, Zn^{2+}. The lead signal was found to be independent of the lead compound used and the solvent type. This is a distinct advantage over flame atomization. Background correction is not found to be necessary.

Reagents and Equipment

The atomic absorption spectrometer and the carbon rod unit (Models AA-5 and 61, respectively) were from Varian-Techtron Pty. Ltd., Melbourne, Victoria, Australia. Sampling was done with a microliter hypodermic syringe (No. 75 SN, The Hamilton Co., Whittier, California). The tip of the syringe was inserted through an injection port of the carbon rod and the source was focused. The sample was dried, ashed, and atomized sequentially by passage of an electric current through the carbon rod. The duration of the atomic absorption signal (and thus the residence time of the atomic population in the optical path) was about 1 sec. Radiation was detected with an R-106 response phototube, and the slit width was 0.100 mm, corresponding to a spectral bandpass of 3.3 Å. The 2833 Å lead line was used.

Chemicals of analytical reagent-grade purity were used without further purification, including lead cyclohexanebutyrate (Pb-CHB) (Eastman Organic Chemicals, Rochester, New York).

Distilled water was passed through an ion exchange column to remove both cations and anions (including lead) before use.

The time and the voltage settings for the dry ash and atomized portions of the furnace cycle were quickly determined and were such that the solvent and the organic matrix were removed before the atomization step, and that the sample was completely atomized and removed from the sample cell with each firing. An adequate rate of argon flow to provide effective sheathing of the carbon rod but not so high as to remove the atomic vapor too quickly from the optical path was about 1 liter/min.

Procedure

Many petroleum samples can be analyzed directly, simply by diluting with xylene in order to reduce the amount of carbonaceous matrix of the oil sample and to give a lead concentration in the appropriate range. In case the carbonaceous matrix and/or the viscosity of the sample are too high, the lead can be extracted by the procedure summarized below.

Weight out a sample of petroleum and dilute with xylene to reduce the viscosity. Add a suitable excess of solid dithizone and extract the lead into a known volume of 40% (v/v) HNO_3. The dilution with xylene is not strictly necessary but facilitates the extraction. Usually the sample size is about 4–10 g and the volume of the aqueous phase 25–50 ml, but these quantities can be increased or decreased to give a lead concentration in the optimal range.

Prepare standard solutions of lead, either from TEL or from Pb-CHB for the analysis of the organolead (organic phase). Use lead nitrate for the analysis of aqueous lead solutions. Blank solutions of xylene of 40% (v/v) nitric acid give no signal, even with full (10×) scale expansion.

Determination of Cobalt, Magnesium, Sodium, Tin, Cadmium, Zinc, and Aluminum in Crude Oils (163)

Comments on the Method

The metals are analyzed by electrothermal atomization. Modifications have been made to the power supply to allow a maximum temperature of 3500°C. However, the highest atomization temperature used is 2300°C. Thus the prospective user of this procedure need not make these alterations.

Both samples and standards are dissolved in xylene. The authors advise that drying, ashing, and atomization temperatures were carefully established and should be adhered to strictly. A calibrated chromel–alumel thermocouple was used for measuring drying and ashing temperatures and an optical pyrometer for atomizing temperature. No problem from nonspecific absorption was noted.

Interferences, enhancement (cobalt and sodium), and depression (magnesium, tin, and aluminum) were found. These were overcome using the standard-additions approach. Zinc and cadmium were free of interference problems and could be analyzed by direct comparison. The accuracy of the method was assessed by comparison with results obtained by a flame atomic absorption. Agreement was satisfactory.

Reagents and Equipment

The instrument was a Varian Techtron Model AA-5 atomic absorption spectrometer, fitted with a Varian-Techtron Model 63 carbon rod atomizer. In order to obtain a faster rate of heating per unit time and a higher final temperature (~3500°C), the power supply to the CRA unit was modified by installing a transformer of larger capability such that a maximum of 24 V across the electrodes (with unloaded graphite tube) was achieved. The maximum current was about 350 A. The power supply unit was equipped with a new triac.

Dry and ash temperatures were measured with a calibrated chromel–alumel thermocouple. Atomization temperatures were measured with an IRCON infrared automatic optical pyrometer (Ircon Co., Niles, Illinois) whose output was fed into a storage oscilloscope (Tektronix, Model 549, Portland, Oregon). Nitrogen or argon was used as a sheath gas.

Samples were introduced into the graphite tube with a 5 μl Hamilton syringe (Model No. 75 N-CH), fitted with a platinum needle and a Chaney adapter. A sample volume of 2 μl was used throughout this study.

A single-element hollow-cathode lamp was used as a narrow line source. An Elektronik 194 (Honeywell) fast-response millivolt recorder was used to record the absorbance signals detected with a Hamamatsu R-106 photomultiplier tube.

Organometallic compounds of known metal content (Eastman Organic Chemicals, Rochester, New York, and Alfa Inorganics, Inc., Beverley, Massachusetts) were used as standards. Dibutyltin-bis(2-ethylhexanoate) was used as a standard for tin. For strontium, magnesium, sodium, cobalt, cadmium, zinc, and aluminum, the cyclohexanebutyrate salts (general formula M-CHB, where M + metal atom and CHG =

cyclohexanebutyrate) were used as standards. Aluminum-2-ethylhexanoate was also used as an additional standard for aluminum.

Fisher certified reagent-grade xylene (Fisher Scientific Co., Ltd., Montreal, Canada) was used as a solvent for tin, cobalt, aluminum, and cadmium. High-purity Aristar-grade xylene (BDH Chemicals Ltd., Poole, England) was used as a solvent for sodium, magnesium, and zinc.

Dibutyltin-bis(2-ethylhexanoate) was dissolved directly in xylene. Na-CHB, Mg-CHB, Cd-CHB, Zn-CHB, Al-CHB, and Al-2-ethylhexanoate were first dissolved in 5 to 7 ml of warm 2-ethylhexanoic acid (Eastman) and the solution then diluted with xylene. All stock solutions contained 500 μg/ml of the metal. Test solutions in the submicrogram/milliliter levels were made from the stock solutions by dilution with xylene immediately prior to the atomic absorption measurement.

The experimental and operating conditions used are shown in Table LV. For each element the optimum time and voltage settings for the drying, ashing, and atomization steps were carefully established. The instrumental settings were chosen such that both the solvent and the organic matrix were removed in the drying and ashing steps. The voltage setting for the atomization step was chosen such that the analyte was completely vaporized with a single firing. The complete vaporiza-

Table LV

Experimental and Operating Conditions

Element	Wavelength (Å)	Lamp current (mA)	Spectral bandpass (nm)	Drying temp[a] (°C) Standard	Drying temp[a] (°C) Oil	Ashing temp (°C) Standard	Ashing temp (°C) Oil	Atomization temp[b] (°C)
Co	2407	6	0.165	85	290	193	247	2300
Mg	2852	3	0.082	75	280	229	297	1100
Na	5890	4	0.165	65	180	180	200	1200
Sn	2863	8	0.165	100	320	220	420	1400
Cd	2288	3	0.165	86	190	169	285	1100
Zn	2138	3.5	0.165	76	340	220	350	1200
Al	3093	12	0.165	86	220	240	260	2200

[a] Measured with a calibrated chromel–alumel thermocouple.
[b] Measured with an infrared automatic optical pyrometer.

tion of the analyte on a single firing was tested by repeating the firing for a second time and recording a zero net absorbance.

Facilities for correction for molecular absorption and/or scattering, by means of a hydrogen hollow-cathode lamp for the ultraviolet region and a tungsten continuum source of radiation for the visible region, were available but were found to be unnecessary under optimum experimental conditions for both the samples and the synthetic standards.

The peak absorbance was measured throughout this study.

Procedure

Calibration curve method. Prepare sample as follows: Accurately weigh samples of crude oil (0.7–1.5 g) into 25 ml volumetric flasks, dissolve in xylene, and dilute with the solvent up to the mark. In the determination of the seven metals in the crude oils analyzed, this method gives accurate results only for cadmium and zinc.

Standard addition method. Prepare the sample as follows. Accurately weigh samples of crude oil (5–12 g) into 25 ml volumetric flasks. Dilute the contents of the flasks to the mark with xylene. Transfer suitable aliquots (typically 2–5 ml) of this solution to a series of 10 ml volumetric flasks containing 0x, 1x, 2x, and 3x amounts of the standard solution of the analyte element, where x stands for the concentration of the element that would give about 0.1 absorbance. Inject a 2 μl sample volume with the syringe into the graphite tube and measure the absorbance with the recorder. The coefficient of variation of the overall determination based on a single sampling is ~3%. However, for better precision, analyses should be done in triplicate and the results averaged. The concentration of the unknown element in the original oil sample is calculated from the standard addition curve.

Determination of Iron, Copper, Nickel, and Manganese in Fats and Oils (164)

Comments on the Method

The procedure may be used for the analysis of hydrogenated oils and crudes. Samples are diluted in methylisobutyl ketone (MIBK) for copper and manganese and MIBK with a small amount of nitric acid for iron and nickel. Xylene was found to be acceptable only for iron and hence was not employed. Grooved graphite furnaces were essential. The ordinary HGA 2000 tubes were unsatisfactory because of the severe spreading of samples even when only 20 μl was used.

The gas interrupt feature of the HGA 2000 gave enhanced absorbances but severe problems due to nonspecific absorption negated its use.

Reagents and Equipment

A Perkin-Elmer 306 atomic absorption spectrometer equipped with an HGA 2000 graphite furnace, deuterium arc background corrector, optimal modification kit, and a Perkin-Elmer 56 strip-chart recorder was used.

Furnace operating conditions used in all determinations are included in Table LVI. The deuterium arc background corrector was used in all determinations.

Perkin-Elmer Intensitron hollow-cathode lamps were used with instrumental parameters shown in Table LVII.

Grooved graphite tubes were used rather than the standard straight tubes. Tube lifetime averaged 150 determinations.

The following Eastman organometallics were used to prepare the organic standard solutions: manganous cyclohexanebutyrate, cyclohexanebutyric acid copper salt, cyclohexanebutyric acid nickel salt, tris(1-phenyl-1,3-butanediono) iron(III). Xylene, MIBK, and nitric acid used in solutions were reagent grade.

Procedure

Prepare 500 μg/ml organic solutions of iron, copper, nickel, and manganese in xylene and ethylhexane acid according to Eastman directions. Dilute these to prepare 100 μg/ml stock solutions. Store in Pyrex glass containers.

The 100 μg/ml standards are convenient for the preparations of the low-concentration working standards required in typical analyses.

Prepare fresh dilute working solutions by diluting in the appropriate solvent solution. Weigh a 1 g sample into an acid-washed 10 ml volumetric flask. Depending on the element of interest, choose the most suitable solvent system to ensure stability for the duration of the analy-

Table LVI

Furnace Operation Conditions

Cycle	Time (sec)	Temperature (°C)
Dry	30	100
Char	20	800
Atomization	20	2300

Table LVII

Instrumental Parameters

Elements	Wavelength (Å)	Slit (Å)
Fe	2483	2
Cu	3247	7
Ni	2320	2
Mn	2795	2

sis. The best solvent systems are MIBK for copper and manganese, and MIBK + 5 drops concentrated nitric acid/10 ml solution for iron and nickel.

Inject 20 μl aliquots of the standard into the furnace, and use the average of three determinations that agree to ±0.005 absorbance units. The concentration of the standard is typically 0.1 μg/ml, since this approaches the upper limit of linearity for iron. Repeat this procedure for a 10 μl aliquot, the equivalent of 0.05 μg/ml. Determine a third smaller aliquot of standard.

Inject 20 μl of the sample and average three determinations agreeing to ±0.005 absorbance units. If the signal produced is beyond the linear portion of the curves presented in the figures, either a smaller aliquot is taken or a smaller sample can be used.

Another method of calibration can be used, the method of additions. Prepare three identical solutions of the sample. To one, add 0.05 ppm of the element of interest with an Eppendorf pipet (5 μl of 100 μg/ml stock solution). To a second, add 10 μl or 1.0 μg/ml. Analyze the three solutions, and record the averages of the determinations.

8
Analysis of Industrial Samples

A wide range of industrial and pharmaceutical chemicals must be analyzed for trace metals. The latter are sometimes health and/or environmental hazards when present in paints, fertilizers, and drugs. In many cases, trace metal contamination of chemicals adversely affects their desired physical and chemical properties.

Relatively high levels of trace substances in laboratory chemicals can negate their use in chemical analysis. Most reagent-grade chemicals have a chemical analysis on the label. However, it is the author's experience, that when high-purity standards are required, these chemicals should be reanalyzed for the elements of interest prior to their use.

Lead in Paint (165)

Comments on the Method

This method avoids a time-consuming ashing step. An MIBK paint or water–paint slurry is used for oil- and water-based paints, respectively. Paint slurry is introduced directly into a Delves cup. The procedure is applied to a wide variety of paints and the results compared to an ashing method. Good agreement is demonstrated between this procedure and one employing ashing. Interferences are minimal with background nonspecific absorption causing the greatest problem. To minimize this effect, no more than 1.0 g of sample should be used. Under these conditions the automatic background corrector can overcome the interference. The detection limit is 5–10 μg Pb/ml.

Reagents and Equipment

A Perkin-Elmer Model 403 atomic absorption spectrometer with a Delves cup accessory, deuterium background corrector, and strip-chart recorder was used for all measurements. The Delves sampling cups, made of nickel, were positioned about 4 mm below the quartz absorption tube aperture. The absorption tube was mounted on a three-slot burner. A somewhat fuel-rich air–acetylene flame was used for sample volatilization. Deionized water was aspirated continuously to prevent the absorption tube from overheating.

A Hamilton 50 μl syringe was used to transfer sample aliquots to the nickel cup. The paint–solvent suspension was stirred by means of a magnetic stirrer equipped with Teflon-coated stirring bars. Ostwald-Folin pipets calibrated to contain (TC) were used for accurate transfer of oil-based standard solutions.

The 2833 Å resonance line with a spectral bandpass of 7 Å was used for all measurements. The source was a lead hollow-cathode lamp operated at a current of 35 mA. Background correction was used for all measurements.

Deionized water was used as a solvent for water-based paints and reagent-grade methylisobutyl ketone (MIBK) was used for oil-based paints. The standard lead solutions in MIBK were prepared from a 5000 μg/ml oil standard, available from Continental Oil Company, Ponca City, Oklahoma. The 1000 μg Pb/ml solution used in preparing standards for water-based paints was obtained from Fisher Scientific Co., Pittsburgh, Pennsylvania.

For analysis of oil-based paints, a 50 μg/ml lead standard solution was prepared by diluting 1.00 ml of the 5000 μg Pb/ml standard to 100 ml with MIBK. Standards containing 2.0, 1.5, 1.0, and 0.5 μg Pb/ml were prepared by transferring, by means of Ostwald-Folin pipet, 4.0, 3.0, 2.0, and 1.0 ml of the 50 μg Pb/ml standard solution to separate 100 ml volumetric flasks. After thoroughly rinsing the pipets with MIBK, the rinsings are transferred to the volumetric flasks and diluted to the mark with MIBK.

For analysis of water-based paints, working standards were prepared by dilution of the 1000 μg Pb/ml stock solution with 1% (v/v) nitric acid solution.

Procedure

After thoroughly mixing the paint sample with a glass stirring rod to ensure homogeneity, transfer an amount of paint containing no more than 0.1 mg of lead to a tared weighing bottle. Weigh the sample to the nearest 0.001 g. Pipet 10 ml of deionized water (for water-based paints) or MIBK (for oil-based paints) into the weighing bottle. Stir the paint–solvent mixture thoroughly with a glass stirring rod, being sure to loosen any bits of paint adhering to the bottom of the weighing bottle. Transfer this slurry to a 250 ml beaker.

Measure exactly 90 ml of the appropriate solvent (water or MIBK) in a graduated cylinder and rinse the weighing bottle with the solvent five times. Transfer the rinsings to the 250 ml beaker. Pour the remainder of the solvent into the beaker. Place a magnetic stirring bar in the beaker and stir at a moderate speed.

After installing the lead hollow-cathode lamp, turning on the deuterium background corrector, adjusting instrumental parameters and mounting the Delves sampling system on the burner chamber, light the flame. Care must be taken to position an empty cup beneath the absorption tube aperture when lighting the flame to reduce the possibility of flashback. After the burner is lit, deionized water must be aspirated continuously to prevent overheating of the absorption tube. The nickel cup should be positioned about 3–4 mm below the absorption tube aperture to obtain optimum precision without sacrificing a great deal of sensitivity. To prevent a possible explosion when shutting down the instrument, the absorption tube and burner head must be allowed to cool before purging the system of residual fuel gas.

Condition a nickel cup by inserting it into the flame for about 20 sec. After cooling, remove the conditioned cup from its holder. By means of a 50 μl syringe, transfer exactly 20 μl of stirred paint slurry to the cup. Evaporate the solvent from the sample holder assembly near the flame or on a hot plate. After drying, return the cup to its holder and insert into the flame. The resultant peak is measured from the volatilized lead on a strip-chart recorder. Repeat this procedure and average the heights of the two peaks. The amount of lead represented by the sample peak is determined by the method of standard additions, whereby 20 μl of a standard lead solution are added to the same nickel cup and the preceding procedure is repeated in duplicate. At least three different concentrations of standard solutions should be run using this method.

Average peak heights vs. standard solution concentration are plotted on linear graph paper. A straight line is drawn between these points and extrapolated to zero peak height. The distance from zero standard concentration at which this line crosses the horizontal axis is a measure of the concentration of lead in the sample slurry. The amount of lead in the paint is calculated from

$$\mu\text{g Pb/ml in paint} = \frac{\mu\text{g Pb/ml in sample slurry} \times 100}{\text{sample weight (g)}}$$

Determination of Trace Metals in Sulfuric and Hydrofluoric Acids and Ammonia (166)

Comments on the Method

Cadmium, iron, copper, lead, manganese, and zinc can be determined in reagent and technical grade acids and ammonia by the following procedure. No lengthy evaporations or addition of reagents are required thus minimizing chance of contamination and time of analysis. Ammonia samples are analyzed by direct injection into the furnace. Sulfuric and hydrofluoric acids are placed in a glassy carbon boat, which is inserted into the furnace. The method of standard additions is used for calibration in the case of both acids. No standard reference samples exist to allow an assessment of accuracy of the method.

Reagents and Equipment

The measurements were made with Perkin-Elmer 303 and 305 atomic absorption spectrometers. A background corrector is not required. Atomizations were made in a Perkin-Elmer HGA 70 or HGA 72 furnace. The analyses were based on measurements of peak heights.

Samples of ammonia solution were transferred directly to the furnace by a 100 μl syringe (Hamilton/Micromeasure) equipped with stainless steel needles. Samples of sulfuric or hydrofluoric acid were transferred with a 5–50 μl adjustable plastic pipet (Finnpipette, Kemistien, Finland) to 28 × 5 × 4 mm glassy carbon boats (Ringsdorff-Werke) specially made to fit into the first of the above types of furnace. The boats were only touched with forceps covered by polytetrafluoroethylene (PTFE) tape.

The samples were evaporated to dryness on a thermostatically regulated hot plate. During evaporations the boats were placed either in a glass Petri dish (for sulfuric acid) or in a similar dish made of PTFE (for hydrochloric acid). A glass or plastic funnel with a diameter somewhat larger than the dishes was placed over the boats, and the funnel was connected to a water jet pump. This setup served the double purpose of removing the acid fumes and reducing the introduction of contaminations from the ambient atmosphere.

The measurements were made at the following wavelengths: cadmium, 2288 Å; copper, 3247 Å; iron, 2483 Å; lead, 2833 Å; manganese, 2795 Å (the relatively high concentrations of manganese sometimes encountered in sulfuric acid were determined by measurements at 4031 Å); and zinc, 2139 Å.

The nitric acid was of Suprapur quality (Merck). The furnace was purged with 99.9% purity argon (by volume). Primary standard solutions were prepared by dissolving the proper amounts of high-purity metals in an excess of nitric acid and diluting with water. Highly diluted secondary standard solutions were prepared just before use, and nitric acid was added to maintain a pH of 2.0.

Procedure

Analysis of sulfuric acid. The trace metals in sulfuric acid were determined by the standard addition technique, two curves being plotted for each metal.

A standard addition curve is obtained by transfer into five of a series of seven preheated boats, 2, 4, 6, 8, or 10 μl of a metal standard solution with a concentration of 50 μg/liter to 1 μg/ml. Transfer the boats to a Petri dish. Place the dish on the hot plate and cover with the funnel. Remove water by evaporation to dryness at about 100°C.

Add a constant volume of sulfuric acid in the range of 20–150 μl (depending on the element to be determined and the concentration) to all boats, and repeat the evaporation at about 180°C. When white fumes are no longer visible, finish the operation by heating for 5 min at about 220°C. The evaporation to dryness of 100 μl of concentrated sulfuric acid takes about 10 min. Remove the dish with the boats from the hot plate, cover, and cool at room temperature.

Finally transfer the boats to the furnace, and atomize the metals by heating for 60 sec at the following temperatures: Cd, 1800°C; Mn, 1900°C; Pb, 1800°C; and Zn, 1700°C. Atomizations are finished by heating for 30 sec at about 1950°C.

The position of each standard addition curve and its intersection with the abscissa is calculated by means of the method of least squares. The calculations can be made with a computer, which can also be programmed to calculate the standard deviation.

All determinations are based on measuring peak areas.

The trace metals in hydrofluoric acid are also determined by the standard addition method; two curves are plotted for each metal.

Establish a standard addition curve by adding 10, 20, 30, or 40 μl of a metal standard solution with a concentration of 10–150 μg/liter to four of a series of six preheated boats. Place the boats in a Petri dish and remove the liquid as described for the analysis of sulfuric acid. Add a constant volume of hydrofluoric acid (10–150 μl is normally suitable) to all boats. Evaporate the acid at about 115°C. Atomize the metals at the temperatures listed above.

All analyses are based upon measurements of the peak areas.

Analysis of ammonia solution. Samples of ammonia solution are evaporated to dryness in the furnace without the use of boats. Heat the furnace to 70–80°C, and add 50–250 μl of the sample by syringe through the radial opening at a slow rate such that a considerable amount is evaporated during the introduction. (This prevents the sample from getting into the cooler end parts of the furnace.)

After evaporation to dryness, heat the sample for 60 sec at about 200°C in order to remove liquid condensed at the end of the furnace. Atomize cadmium and lead as described above, but atomize copper and iron at 2500°C (in furnace HGA 70) and 2540°C (in furnace HGA 72), respectively.

The method was calibrated by using constant volumes of suitable metal standard solutions.

Copper and iron are determined by measuring peak heights, and the other two by measuring peak areas.

Determination of Iron, Copper, Nickel, Manganese, Chromium, and Cobalt in High-Purity Glass (167)

Comments on the Method

The metals are extracted into MIBK at pH 6 using diethyldithiocarbamate, which negates interference problems from the major matrix constituents. Analysis is done in a graphite furnace. Great care is essential to avoid adding trace impurities from reagents and containers. The authors emphasize the following important points:

(1) All plastic and glassware should initially be thoroughly cleaned and then retained solely for this type of work.

(2) Reagents of the highest purity available should be used and where possible further purification undertaken if necessary.

(3) The absolute minimum of apparatus should be used.

(4) The samples should be dissolved under an extraction hood and all work should be carried out in a clean atmosphere, preferably in a clean air cabinet.

The procedure gives low results for chromium due to losses as chromyl chloride and hence data for this element should be used only to give a rough idea of the concentration. Good recoveries, assessed by analyzing glasses with known impurity levels, are obtained for the other elements.

Reagents and Equipment

Standard solutions containing 100 μg/ml of the element being determined were prepared as follows:

Copper and nickel. Dissolve the pure metals in the minimal quantity of nitric acid and then dilute appropriately with distilled water.

Manganese and iron. Dissolve their sulfates in the appropriate quantity of distilled water, containing 1 ml of hydrochloric acid per liter.

Cobalt. Dissolve cobalt chloride in the appropriate quantity of distilled water, containing 1 ml of hydrochloric acid per liter.

Chromium. Dissolve potassium dichromate in the appropriate quantity of distilled water.

Composite standard solutions containing all six elements, required for calibration purposes, were prepared as required by the appropriate dilution of the stock solutions with distilled water.

"Suprapur" hydrofluoric and perchloric acids (Merck, Darmstadt) were used for dissolving the samples. All other reagents used were of analytical-grade quality.

A Perkin-Elmer Model 306 atomic absorption spectrometer was used with an HGA 70 graphite furnace, and a Model 165 recorder. Argon was used to provide the inert atmosphere within the graphite furnace. Grooved graphite tubes were used for all determinations; these were found to remove the problems invariably encountered when the normal graphite tubes are used for the analysis of organic solvents.

Oxford Sampler micropipets fitted with disposable plastic sampling tips were used for introducing solutions into the graphite tube.

Procedure

All samples are analyzed in duplicate where possible and duplicate blank determinations are done simultaneously.

Place the glass samples in a platinum dish and heat in a furnace at 800–900° for 0.5–2 min (depending on the size and shape of the sample), then shatter them by dropping a few milliliters of distilled water onto the pieces of glass. Grind the samples to a fine powder with an agate pestle and mortar. Weigh 1.00 g of the sample into a platinum crucible, add 4 ml of hydrofluoric and 5 ml of perchloric acids, and heat gently to dryness. For carbonates, weigh 1.00 g of sample into a platinum crucible, add 3 ml of perchloric acid to dissolve the sample, and heat gently to dryness.

Both types of sample are now treated in the same way. Dissolve the residue in a mixture of 0.1 ml of perchloric acid and 4 ml of water and transfer the solution to a 10 ml volumetric flask. Wash out the crucible with 5 ml of a solution containing about 5% (w/v) sodium diethyl-dithiocarbamate and 10% (w/v) sodium acetate, buffered at pH 6 and previously purified by extraction with MIBK. Transfer these washings to the volumetric flask and allow the flask and contents to stand for 15 min. Add 1 ml of MIBK to the flask and shake thoroughly for 3 min, and then allow the layer of MIBK to separate in the neck of the flask; this extract is used for all subsequent determinations.

Determine the concentrations of iron, copper, nickel, cobalt, manganese, and chromium in each sample by flameless atomic-absorption spectrometry under the instrumental conditions shown in Table LVIII.

Table LVIII

Instrumental Conditions for the Determination of Iron, Copper, Nickel, Cobalt, Manganese, and Chromium by Flameless Atomic Absorption Spectrometry

	Fe	Cu	Ni	Co	Mn	Cr
Sample aliquot (μl)	20	20	50	50	20	20
Program setting[a]	6	6	7	7	6	6
Atomization voltage[a]	9.5	9.5	9.5	9.5	9.5	9.5
Drying time (sec)	30	30	90	90	30	30
Ashing time (sec)	30	30	30	30	30	30
Atomization time (sec)	5	5	5	5	5	5
Scale expansion	2×	2×	10×	5×	1×	2×
Calibration range (μg/ml)	0.02–5.00	0.01–0.50	0.02–0.50	0.01–0.50	0.01–0.50	0.01–0.50
Wavelength (Å)	2483, 3720	3247	2320	2407	2795	3579

[a] These figures refer to the use of grooved-type graphite tubes.

Establish the concentrations from calibration graphs constructed using aqueous standards solutions and instrumental conditions identical to those used for the samples. Then determine the true sample concentrations by subtracting the blank readings from the sample readings.

Determination of Mercury in Mercury-Containing Pharmaceuticals (168)

Comment on the Method

A digestion of the sample in concentrated hydrochloric acid is used. The fact that no oxidizing reagent is required is surprising. Mercury losses have been reported for many other sample types under these conditions. However, these authors demonstrate 98.9–103.5% recoveries of mercury for 10 mercurial compounds.

A cold vapor absorption method is employed. No interferences, including nonspecific absorption, were encountered. The author, however, recommends background correction be used as a precaution.

Reagents and Equipment

An atomic absorption spectrometer equipped with a mercury hollow-cathode lamp, a hydrogen continuum lamp to provide background correction capacity, and a cylindrical flow-through cell (2.1 cm i.d. × 10.0 cm) with quartz windows were used. Absorbance measurements were monitored with a 0–100 mV strip-chart recorder. Mercury analysis was performed under the following conditions: wavelength, 2537 Å; hollow-cathode lamp current, 3 mA; scale, 0.5; circulating pump flow rate, 4.8 liters; slit width, 160 μm; and chart speed, 2.5 cm (1 in.)/min.

The reduction vessel consisted of a 250-ml two-necked distilling flask with a vertical side neck and 24/40 joints. The reducing solution was added via a 100 ml cylindrical separator attached to the side neck of the reduction vessel. All connections within the system were of Tygon tubing (5.0 mm i.d.).

Glassware was thoroughly rinsed with hot nitric acid (50% v/v) followed by distilled water prior to use. Dilutions of samples, pure compounds, and mercury standards were carried out in the presence of hydrochloric acid to minimize absorption of mercury ion on glassware surfaces.

Crystalline mercuric chloride prepared in an acidic medium was employed as the standard mercury solution. All other chemicals and reagents were reagent-grade, commercially available materials and were used without further purification.

A standard stock solution was prepared by dissolving 0.1354 g of crystalline mercuric chloride in 1 N HCl and diluting to 100.0 ml with the same solvent. This solution was prepared fresh on a biweekly basis. Suitable dilutions were prepared in 0.05 N HCl to provide a working standard solution having a concentration of 0.25 μg Hg/ml. These dilutions were carried out daily and just prior to the quantitative reduction step.

Procedure

Treat accurately weighed portions of solid materials or measured aliquots of the liquid forms representing the various mercurial compounds or respective products under one of the following heating conditions with occasional swirling of the mixture:

(A) Heat for 1 hr on a steam bath with 75 ml of concentrated hydrochloric acid in a 1 liter volumetric flask.

(B) Boil gently for 25 min on a hot plate with 75 ml of concentrated hydrochloric acid in a 300 ml Erlenmeyer flask containing five glass beads and fitted with a 6.5-cm-diameter powder funnel. Take care to prevent the solution from going to dryness by addition of 5–10 ml of the acid when necessary.

(C) Heat for 1 hr on a steam bath with 75 ml of diluted hydrochloric acid–nitric acid mixture in a 1 liter volumetric flask.

(D) Heat for 15 min on a steam bath with 80 ml of dilute hydrochloric acid (10% v/v) in a 1 liter volumetric flask.

The required heating condition for each mercurial compound and the respective commercial preparation is noted in Table LIX. Upon completion of the heating step, cool the 1 liter flask and contents under tap water to room temperature and dilute the acidic solution to volume with distilled water. When heating condition B is employed, the cooled acidic solution is quantitatively transferred to a 1 liter volumetric flask with distilled water and diluted to volume.

Prepare further dilutions with 0.05 N hydrochloric acid to provide an assay solution having a final concentration of approximately 0.25 μg Hg/ml. Transfer 3 ml of this assay solution to the reduction vessel. The system is normally equilibrated with several standard determinations prior to the reduction of sample aliquots. A reagent blank consisting of the appropriate acid solution should be carried through the entire procedure in a similar manner to the samples. Quantitation is accomplished by direct comparison of the sample absorbance value to the absorbance obtained with 3.0 ml of the working standard solution of mercury.

Table LIX

Heating Conditions

Compound	Heating condition
Thimerosal	C
Mersalyl	A
Meralluride	A
Mercaptomerin sodium	C
Phenylmercuric acetate	B
Chlormerodrin	A
Merbromin	A
Nitromersol	B
Phenylmercuric nitrate	B
Mercuric oxide, yellow	D
Ammoniated mercury	D

Bulk drug substances or reference compounds. An accurately weighed portion of the powdered material, equivalent to 60 mg of mercury is taken for analysis. After completion of the appropriate heating step and dilution of the acid solution to 1 liter with distilled water, prepare an assay solution by further dilution of 2.0–500.0 ml with 0.05 N hydrochloric acid. Transfer a 3.0 ml aliquot of the assay solution to the reduction vessel for quantitative purposes.

Tablets and gels. An accurately weighed portion of a powdered tablet composite or well-mixed gel preparation is taken for the analysis. The general procedure is followed.

Solutions and tinctures. An accurately measured aliquot is evaporated to dryness under a current of air on a steam bath, the required acid is added, and heating condition A, B, or C was followed (general procedure).

Injectables. An accurately measured aliquot is taken directly for the analysis as described under procedures.

Ointments. An accurately weighed portion of the well-mixed product is transferred to a 125 ml separator and dispersed by shaking with 50 ml of ether. The ethereal dispersion is extracted four times with 20 ml volumes of dilute hydrochloric acid (10% v/v), and the acidic extracts are drained into a 1 liter volumetric flask. The general procedure is followed, beginning with heating condition D. This condition is employed to facilitate the removal of the residual ether present.

Calculations. The amount in milligrams of the mercurial compound present in the weighed portion or aliquot is calculated in the following manner:

$$mg = \frac{A_u}{A_s} \times C \times D \times \frac{E}{200.59} \times \frac{1 \text{ mg}}{1000 \text{ } \mu g}$$

where A_u and A_s represent the absorbance values obtained for the reduced sample and standard aliquots, respectively (corrected for any reagent blank absorbance); C is the concentration (μg Hg/ml) in the final diluted mercuric chloride standard solution; D is the appropriate dilution factor; E is the molecular weight of the mercurial compound; and the value of 200.59 represents the atomic weight of mercury. Calculation of the quantity of the mercurial compound per dosage unit or labeled amount is achieved by introduction of the appropriate factors into the equation.

Linearity. Under the assay conditions described, a linear relationship between absorbance and micrograms of mercury is obtained over a 0–1.4 μg range. A quantity of mercury equal to 0.75 μg is selected for the reduction step of the general procedure.

Determination of Arsenic, Selenium, and a Number of Heavy Metals in Fertilizers (169)

Comments on the Method

The following methods were developed for sewage sludge-type fertilizers. However, experience has shown that many of the procedures are applicable to a range of highly organic to inorganic fertilizers.

The decomposition methods used are for strong acid-extractable metals. These and not the silicate-bound metals are normally of interest. Should the latter be required, a hydrofluoric acid treatment may be used in addition to the steps in the procedures. This negates the determination of arsenic, selenium, and mercury. For the former two elements, a fusion using NaOH will result in complete recovery (see also general section of hydride determination).

The methods given below have been tested on Canada Center for Inland Waters, Standard Reference Sewage Sludges, with good results.

Reagents and Equipment

An Instrumentation Laboratories Model 153 equipped with a high solids burner was used for flame work.

Mercury analyses were done with the Coleman MSA 50 analyzer or a Perkin-Elmer 305B unit modified with a cold-vapor tube. A deuterium arc background corrector was used. For furnace work a Perkin-Elmer 303 was equipped with an HGA 2000 furnace.

All acids and chemicals used were reagent grade, Fisher Certified. Blanks were run at all times to eliminate problems from contamination.

Standard aqueous metal solutions (1000 μg/ml metal) were prepared from reagent grade metals or metal salts. Working solutions were prepared by diluting the above an appropriate amount being careful to maintain an acidity (nitric acid) of approximately 5% by volume.

Procedure (Chromium, Vanadium, Cadmium, Lead, Nickel, Zinc, Iron, Copper, Manganese)

Weigh a 0.5 g portion of a representative sample into an acid-washed (aqua regia) 100 ml beaker. Add 9 ml hydrochloric and 3 ml nitric acids. Heat the sample on medium heat for 1 hr. (During this time samples evaporate to near dryness.) If any samples go to dryness, add 1 ml of nitric acid. Dilute the samples to 10 ml with water and filter (Whatman No. 42) into a 25 ml flask. Wash the filter well with water and dilute the sample to volume.

For sludge and sludge-based fertilizer samples the following dilutions are commonly needed. Dilute 2 ml of stock solution to 25, 100, and 250 ml. Dilute 2 ml of the above 25 ml dilution to 250 ml. The acid content is made 5% with nitric acid in each case.

Determine the solutions on the atomic absorption using the conditions outlined in Table LX. Background correction is done using either the lines listed in Table LX or the deuterium arc corrector when available.

Standard working solutions are determined with each sample set to obtain calibration curves. In addition, a Canada Center for Inland Waters Reference Standard Sewage Sludge (Burlington, Ontario, Canada) is run with each batch.

Mercury. Weigh a 0.1–1.0 g sample, depending on mercury content, into a BOD bottle supplied with the Coleman MAS 50. Add 15 ml of 7 N nitric acid, 15 ml 1:1 sulfuric acid, and 0.5 g of potassium persulfate. Heat on a low heat, less than 75°C, for 15 min. Cool to room temperature and add 10 ml of 0.1% hydroxylamine hydrochloride fol-

Table LX

Atomic Absorption Parameters

Element	Analytical line (Å)	Background line (Å)	Oxidant	Slit (μm)
V	2184	3196	N_2O	80
Cr	3579	(3660)[a]	Air	40
Cd	2288	2264	Air	80
Pb	2833	2820	Air	80
Ni	2320	2316	Air	40
Zn	2139	2120	Air	80
Cu	3247	2961	Air	80
Mn	2795	2817	Air	40
Fe	2483	not needed	Air	40

[a] Line found suitable for our lamp—accurate wavelength not known.

lowed by 10 ml of 10% stannous chloride solution and run on the Coleman MAS 50 analyzer or PE 305B. The above procedure is modified from that recommended by Hatch and Ott for other sample types.

Silver. For samples that are low in silver (usually 5 μg/g or less) the above aqua regia procedure may be applicable. However, for many samples silver is precipitated during the above procedure as insoluble silver chloride. If this occurs the following procedure is recommended.

Place 0.5 g of sample into a 100 ml beaker. Ash the sample at 450°C to a light ash and extract the ash over medium heat of a hot plate with 9 ml of nitric acid.

Using the Perkin-Elmer 303 fitted with the HGA 2000 furnace the conditions in Table LXI are used. Allow the instrument to warm up for at least 20 min to stabilize the deuterium arc. Align the arc and silver lamp beams very carefully.

Table LXI

Experimental Conditions for Silver

Absorbing line	3281 Å
Background correction	Deuterium arc and/or 3324 Å nonabsorbing line (see accompanying discussion)
Slit	7 Å
Dry (100°C)	30 sec
Ash (1000°C)	30 sec
Atomize (2600°C)	8 sec

Pipet 5–50 μl aliquots of standard and sample solutions separately into the furnace. Acitivate the thermal program. A standard curve is produced by reading peak heights of standards. Unknowns (sludges) are read off the standard curve.

As indicated below, the background corrector gives error signals from time to time. The nonabsorbing-line method must be used in these cases as follows:

The nonabsorbing line at 3324 Å is used. The samples are run under the same conditions as for the absorbing line. The background signal thus obtained is subtracted.

Molybdenum and vanadium.   Using the Perkin-Elmer 303 and HGA 2000, the conditions given in Table LXII are used.

Allow the instrument to warm up for at least 20 minutes to stabilize the deuterium arc. Carefully align the deuterium arc and hollow-cathode lamp beams. Analyze the samples by the method of standard additions. Use 50 μl aliquots.

Determination of Molybdenum, Cobalt, and Nickel in Hydrodesulfurization Catalysts (170)

Comment on the Method

Flame atomic absorption is used to analyze hydrodesulfurization catalysts for molybdenum, cobalt, and nickel. Nitrous oxide–acetylene is used for molybdenum. The only serious interference in the catalyst matrix is aluminum and compensation is made by adding this element to the standards. Good agreement with known values was obtained for the analysis of manufacturers standards by the proposed procedure.

Table LXII

Experimental Conditions for Molybdenum and Vanadium

Absorbing line	
Mo	3133 Å
V	3184 Å
Background correction	Deuterium arc
Slit	7 Å
Dry (100°C)	60 sec
Ash (1200°C)	40 sec
Atomize (max temp.)	30 sec
Purge gas (N_2)	2 ft³/hr

Reagents and Equipment

A Varian Techtron AA5 atomic absorption spectrometer with the source and detector modulated at 258 Hz was employed. The hollow-cathode lamps (molybdenum, cobalt, and nickel) were all Varian single-element, high-intensity sources. The detector was equipped with an R446 photomultiplier. For the air–acetylene flame analysis, a 10 cm slot burner (Varian) head was used while a 5 cm-slot burner head (Varian) was employed for nitrous oxide–acetylene analysis.

In all cases the source and burners were allowed to warm up for 15 min before the analyses were performed. The wavelength selection was always peaked about the wavelength given in Table LXIII for maximum sensitivity. While the solution is still hot it is quantitatively transferred to a 100 ml volumetric flask. It is then cooled and diluted to volume.

Samples were diluted prior to determination to the same concentration range of the standards. Typical dilutions were 1/10 for molybdenum, 1/4 for cobalt, and 1/20 for nickel.

Calibration curves from molybdenum, cobalt, and nickel standards were constructed with solutions containing 400 μg/ml alumina.

To ensure that the results were not affected by the burner becoming slightly clogged because of the high aluminum concentration in the nitrous oxide–acetylene flame the following sample–standard sequence was used: U1-S1-U1-U2-S2-U2-U3-S3-U3-U4-S4-U4-U5-S1-U5-U6-U7-S3-U7-U8-S4-U8, where U is the unknown and S is the standard. From these readings, absorbances vs. concentrations plots were constructed.

Samples are analyzed as follows, by the method of standard additions. Three solutions are used, the sample by itself and then two aliquots of the sample spiked with an appropriate concentration of dilute standard solution (usually sample + 0.2–0.8 μg/ml).

Table LXIII

Operating Parameters for Catalyst Analysis

Parameter	Mo	Co	Ni
Wavelength (Å)	3133	2407	2320
Slit width (μm)	100	25	50
Lamp current (mA)	5	5	5
C_2H_2 flow (liters/min)	4	4	0.75
N_2O flow (liters/min)	9	10	—
Air flow (liters/min)	—	—	10.5
Solution uptake (ml/min)	3.1	3.1	3.1

Procedure

Weigh accurately about 0.5 g of sample (at the same time weigh another aliquot for moisture content determination) into a 125 ml Teflon beaker. Add 20 ml of concentrated sulfuric acid and treat dropwise with 10 drops of concentrated hydrofluoric acid. When the reaction subsides, add another 5.0 ml of concentrated hydrofluoric acid.

The beaker is heated for 45 min at 90°C; if the solution is not clear at this time (due to a high content of SiO_2) another 10 ml of concentrated hydrofluoric acid is added. When the solution is clear it is heated for a period of 1–2 hr at 60–80°C to ensure the complete loss of the residual hydrofluoric acid. Then 20 ml of distilled water is added and the solution is heated to 80°C. Place 50 μl aliquots of these solutions separately in the furnace and run using the conditions listed above.

9
Determination of Metal Compounds

Total metal data abound and yet most of the important metal-related problems facing environmental, clinical, biochemical, and agricultural scientists have not been solved. The pollution incidents involving mercury have shown that total metal data are insufficient and often misleading in assessing the environmental and health hazard of this metal. Metal compounds (species) must be identified and determined quantitatively.

Metal speciation is, at best, a difficult task. It is therefore surprising how little use has been made of atomic absorption spectroscopy as a detector in metal speciation apparatus. Atomic absorption is metal specific and hence will respond only to compounds containing the analyte metal, thus yielding output of the simplest possible form. Interfacing this detector with most separatory equipment, e.g., chromatography, is extremely simple. Laboratories engaged in metal analyses are usually equipped with atomic absorption instrumentation and hence are within easy reach of expanding their capabilities into the metal speciation field.

Kolb et al. (171) in 1966 first used atomic absorption spectroscopy for metal speciation studies. Effluent from a gas chromatographic column was transferred to the nebulizer using a short piece of heated metal tubing. In 1973, Manahan and Jones (172) reported the interfacing of atomic absorption spectroscopy to high-pressure liquid chromatography. The column effluent was connected directly to the burner through the nebulizer capillary. Despite these and other pioneering studies that demonstrated the usefulness of the approach, there have only been, up until 1977, about 50 literature references to metal speciation studies using atomic absorption as the detector (173).

Selective Volatilization (174)

Comment on the Method

It is frequently important to study the loss by volatilization on drying and ashing of some of the more volatile metals as metals and compounds. These include mercury, cadmium, zinc, etc. This can be done simply in a quartz T tube, where the volatilizer and atomizer furnaces are separated. It is crucial in this application to use background correction to avoid nonspecific signals caused by smokes and other nonanalyte vapors.

Reagents and Equipment

A Perkin-Elmer 305B atomic absorption unit with deuterium arc background corrector was used. Hollow-cathode lamps were recent models available from Westinghouse.

The quartz T tube furnace is shown in Fig. 42, together with a coil chromatographic column not needed in this application. All tubing is quartz, but the male section of the joint may be glass if desired.

The volatilization section is 20 cm long × 4.5 cm i.d., narrowing to a neck 8 cm long × 8 mm i.d., which is sealed to the center of the atomizer T tube. The atomizer is 10 cm long × 2 cm i.d. The gas inlet tubes in the neck and on the male part of the ground quartz joint inlet are 6 mm i.d. The volatilization and atomizer-heated sections are wound with gauge 20, Chromel C wire with a resistance of 0.65 Ω/ft. Asbestos insulation covers the windings to a 1 cm thickness. The windings are connected to variacs. A temperature of 900–1000°C is used in the atomizer. The hydrogen flow rate is 1 liter/min.

The nitrogen tube on the volatilizer neck, opening directly into the neck, is used with the gas chromatographic column, nitrogen is flushed throughout this section using an inlet tube in the male joint. In both cases the flow rate should be 6 liter/min.

A Sargent Welch Model DSRG double-pen recorder was used. One channel recorded the temperature of the volatilization chamber, by connection to a platinum–rhodium thermocouple and cold reference junction. The other channel monitored the atomic absorption signal, in absorbance.

The T tube atomizer is placed on the burner as in Fig. 43. While monitoring the hollow-cathode lamp beam, the atomizer is moved up and down, in and out, with the burner adjustment screws until minimum absorbance is noted.

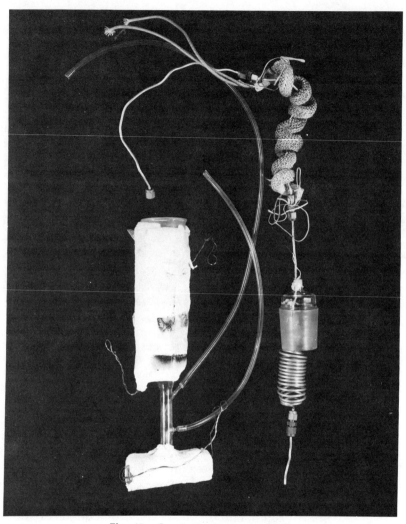

Fig. 42. Quartz T tube furnace (174).

Procedure

Selective volatilization from solids or liquids. Warm up the atomizer atomic absorption unit, deuterium arc, and recorder for 20 min. Set instrument parameters according to the manufacturer's instructions. It is important that the deuterium arc beam be properly aligned with the hollow-cathode beam. Place a sample—frozen, room temperature, etc., as required—into the center of the volatilizer. While

Fig. 43. Quartz T tube furnace mounted on the atomic absorption unit (174).

connecting the male section of the ground glass joint, adjust the position of thermocouple to be close to or touching the sample. Initiate the nitrogen flow. After 10–20 sec, turn on the hydrogen flow to the atomizer. Zero the atomic absorption unit and the activate temperature program. Monitor temperature and absorbance.

Temperature calibration.   Place the thermocouple in a suitable sample of known melting point in the volatilizer. Using the method for selective volatilization, increase the temperature until melting occurs. At this point a sudden aberration is noted on the temperature record. In addition, an absorbance signal is commonly recorded as a significant vapor pressure of the compound occurs. Several compounds of known melting point, covering the temperature range of interest, should be run.

Determination of Selenium Alkanes Transpired by Plants (175)

Comment on the Method

A preconcentration using an absorption column is necessary to bring the level of the selenium compounds to within the detection limit of atomic absorption. A quartz tube atomizer is used in place of a flame to minimize absorption of the 1960 Å resonance wavelength of selenium by the flame gases. The method, although only demonstrated for selenium alkanes, would be applicable to any selenium compounds volatilizable under the conditions listed below.

Reagents and Equipment

A quartz tube atomizer as described above in the selective volatilization procedure is used.

Hydrogen gas is injected into the atomizer using the gas inlet quartz tube closest to the atomizer. This tube runs through the neck to a point 5 mm inside the atomizer. The hydrogen flow used is 1 liter/min. The nitrogen inlet tube opens directly into the neck. Nitrogen flow used was 6 liters/min.

The columns were constructed by Chromatographic Specialties, Brockville, Ontario, in a helix-wound form (Fig. 42). The smallest diameter that could be wound without causing flattening and hence obstruction of the flow was 4 cm. The column was constructed from a 122 cm length of 3-mm-i.d. aluminum tubing. A length of empty aluminum tube was left on both ends of the column. The outlet was long enough to reach to within 5 mm of the inside of the atomizer. A 20% polymetaphenylether (5 ring) on 60/80-mesh Chromasorb W column was found to be suitable for selenium work. This column is stable to 250°C.

The preabsorption column was 35 cm long and similar in other dimensions to the above main column. Silicone oil DC550 on 60/80-mesh Chromasorb W was used as a packing material. After the collection period was over, the precolumn, wrapped in heating tape, was connected to the main column using Swagelock fittings.

Columns must be cured for several hours at 200°C with the carrier gas on, connected to a calibrated Pt/Rh thermocouple and cold reference junction. Thus, temperature can be plotted if desired. The other channel monitored the atomic absorption signal in absorbance.

Reagents were obtained from Alfa Products Ltd., Beverley, Massachusetts. The selenium alkanes, dimethyl selenide, diethyl selenide, and dimethyl diselenide were shipped as liquids sealed in ampules. Working standards of the selenium compounds were prepared by diluting to an appropriate level in 95% ethanol.

The T-tube atomizer is placed on the burner as in Fig. 43. While monitoring the signal from the hollow-cathode lamp, the atomizer is moved up and down, in and out, with the burner adjustment screws until minimum absorbance is noted.

Astragalus, racemosus plants, were grown in Hoagland nutrient solution. Selenium was added as Na_2SeO_4 in the 10–20 μg/ml amounts. Plants were used at the age of 3–6 weeks.

Procedure

Volatile selenium compounds are released from plants and standards using heat.

Place plants or standards in a U tube that is connected to the precolumns by plastic tubing. Surround the precolumn with dry ice and connect to a vacuum pump. Apply low heat to the U tube while drawing air through at a slow rate (15 ml/min) using the vacuum pump.

Connect the precolumn to the main column while maintaining the former in dry ice. Connect a nitrogen cylinder to the other end of the precolumn and move the dry ice to the main column. Wind heating tape around the precolumn and elute the selenium compounds onto the main column using a nitrogen gas flow rate of 15 ml/min and an elution temperature of 175°C. Elution requires 10 min.

During this interval, turn on the atomizer to warm up (20 min). The atomic absorption unit, deuterium arc and recorder should also be stabilizing during this interval. Use the instrument parameters recommended by the manufacturer. It is crucial that the deuterium arc beam be properly aligned with the hollow-cathode beam. Adjust the nitrogen flow to the atomizer to 6 liters/min.

Insert the column into the cool column chamber of the T tube. Initiate the H_2 flow to the atomizer, at 1 liter/min. If a temperature plot is desired, inset the thermocouple so the tip rests on the column. Turn on the N_2 flow through the column at 15 ml/min and initiate the heat program to the column by adjusting the variac to a reading that will give the desired heating rate (e.g., 4°C/min). After the desired maximum temperature has been reached, in this case 135°C, discontinue the procedure.

Determination of Mercury Compounds (176)

Comment on the Method

Fish tissue may be analyzed for alkyl mercury compounds by this procedure. This type of analysis is crucially important to the assessment of the hazard posed by mercury in the environment.

Organomercury compounds are extracted with HCl and benzene. An aliquot of benzene is injected into the gas chromatograph.

Reagents and Equipment

Benzene. Purity is such that injection of 10 μl of a 100 ml aliquot concentrated to 10 ml does not produce a detectable signal by electron capture.

Hydrochloric acid. Ultrahigh-purity hydrochloric acid (Ulrex, J. T. Baker Co.) is extracted several times with equal volumes of benzene until injections of the organic layer do not show peaks eluting after the solvent front employing an electron capture detector system.

Atomic absorption spectrometer. A Perkin-Elmer Model 303, equipped with a 10 cm quartz cell and a Servo/Riter II recorder was operated under the conditions recommended by the manufacturer.

Flameless mercury analyzer. A Coleman Instruments Model MAS-50 equipped with a 2537 Å narrow bandpass filter is used. The cell is disconnected from the air pump. One side of the cell coupled to the exit of the combustion furnace by means of Teflon tubing (1.5 mm i.d.). The other side is directed to an activated charcoal trap.

Combustion furnace. A Dohrman Instrument Co., Model S-100, fitted with a 31 × 0.8 cm quartz tube is maintained at 780°C with an oxygen gas flow of 25 ml/min. Inlet temperature is 280°C.

Gas chromatograph. Microtek, Model MT-220, equipped with an electron capture detector system, a flame photometric detector Melpar Inc., Model 100 A.T., photomultiplier tube, type 9526 mercury filter, 253.7 nm, and a U-shaped glass column 0.25 in. × 6 ft packed with 5% HIEFF-2AP on Chromosorb WHP, 80/100 mesh (Applied Science Lab, State College, Pennsylvania). Temperatures: inlet, 200°C; column, 170°C; detector, 200°C; transfer lines and outlet cock, 200°C; column flow (N_2), 120 ml/min. PPD flame: O_2 and H_2 are mixed in different proportions as described in the results and discussion.

Homogenizer. Sorval Omni Mixer (Ivan Sorval Company, Norwalk, Connecticut).

Procedure

Place a sample of fish tissue in the homogenizer cup and macerate for 5 min at a setting of 4.5. Weigh one of the homogenized tissues into a 15 ml glass-stoppered centrifuge tube. And to the sample tube 1 ml of water and 1 ml of concentrated HCl. Mix the sample on a Vortex mixer for 3 min. Add 2 ml of benzene, shake the tube vigorously for 5 min, and centrifuge. Remove the benzene layer, dry over anhydrous sodium sulfate, and inject a suitable aliquot into the gas chromatograph.

After injection of the dried benzene extract, open the vent to the combustion furnace unit, thus preventing the solvent from entering the furnace. As soon as the benzene has eluted, close the vent and record the mercury absorption.

A calibration curve is run with each set of samples by substituting water for tissue and using identical volumes of reagents and benzene.

Determination of Hexavalent Chromium in Waters (177)

Comment on the Method

Hexavalent chromium has been shown to be carcinogenic at very low concentrations, levels above 50 μg/liter being hazardous. Trivalent chromium, on the other hand, is not harmful except at relatively high concentration. It is meaningless to analyze for total chromium in drinking waters. The following method, involving anion exchange, allows a preconcentration of hexavalent chromium to a level detectable by atomic absorption. Trivalent chromium is rejected in the effluent. When sufficient hexavalent chromium has been accumulated on the column, the column outlet is connected directly to the nebulizer of the burner (Fig. 44). The chromium is eluted at 6 ml/min, which is compatible with the nebulizer flow rate.

Reagents and Equipment

As shown in Fig. 44, 10 ml burets are modified. A nebulizer capillary is cemented in a female glass joint as shown. Two ml of glass beads are placed in the bottom, which help increase the reproducibility of the hexavalent chromium peaks. The columns are packed with 2 ml of Baker ANGA-316, 16–50 mesh, weakly basic anion exchange resin in the chloride form. Any atomic absorption unit can be employed. Best results are obtained when a peak area mode is used.

Procedure

Equilibrate the column with 1 M sodium chloride solution. Pass water samples (pH 6.0–6.5) through the column at 5–10 ml/min. Sufficient sample is used to give a detectable quantity of hexavalent chromium. For example, 150 ml of a 10 μg/liter solution gives a peak

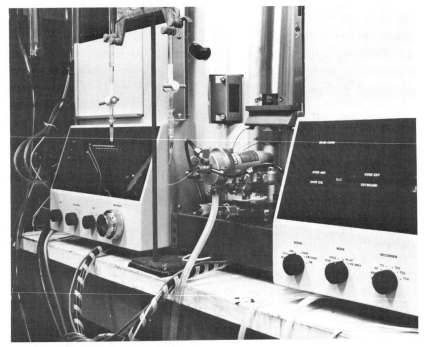

Fig. 44. Buret ion exchange column interfaced with atomic absorption for the analysis of hexavalent chromium (177).

twice the detection limit. The liquid level should be allowed to drop to the top of the resin bed. Add approximately 1 ml of eluent, a 75% saturated solution of ammonia, to the column. Again allow the liquid level to drop to the top of the resin bed. Place the female glass joint on the column. Allow a reaction period of 1 hr. Attach the capillary to the nebulizer. Fill the column with eluent and open the stopcock. Use an integration period of 60 sec, the beginning of the period coinciding with the opening of the stopcock. Aspirate blank eluent between column runs to allow zero adjustment.

To maximize throughput per sample, 10–20 columns are prepared in one batch.

Determination of Individual and Total Lead Alkyls in Gasoline (178)

Comment on the Method

Very simple gas-chromatographic instrumentation is used. A suitable gas-chromatographic column is wrapped in heating tape and interfaced with the flame atomizer through the side of the burner head (Fig. 44). Work in our laboratory also confirms that this very simple approach is acceptable when using atomic absorption spectroscopy as the detector for chromatography, because of the metal specificity of the former technique.

Reagents and Equipment

The column, 3 ft × 0.25 in. o.d. (3/16 in. i.d.) stainless steel is packed with 10% PEG 20 *M* (Carbowax) on 100–120 mesh Porasil C (Water Associates).

The packing is prepared by first dehydrating the support by heating at 150°C for 2 hr under vacuum. A solution of 10% PEG 20 *M* in water-free methanol is added and stirred in, and the solvent is then removed by warming under vacuum in a rotating flask.

The column is packed under vacuum using vibration to compact the packing, until no further settling occurs, then plugged. At the injection end 1.5 in. of free space are left. The column is then conditioned at 250°C for 2 hr.

The use of spherical silica beads as support allows a uniform high column packing density, which gives fast, high-resolution separations from relatively short columns. A polar stationary phase is used because it is easier to coat onto Porasil, giving good peak shapes; also, TML is retained sufficiently so that its peak is not too sharp, as the response of the atomic absorption detector is best suited to broader peaks, while TEL is still eluted in a reasonable time in spite of the large difference in their boiling points.

The column is coiled and wound with a 4-ft-long 240 W heating cord. The windings must be evenly distributed (uneven windings cause peak tailing) except at the injection point, where twice the number are wound on for about an inch. The column is coupled to a short length of $\frac{1}{16}$-in.-o.d. stainless steel tubing, which connects onto the AAS burner as shown in Fig. 45. This tube is also wound with heating cord.

The cord is powered from a variable transformer set to around 60%, which maintains the column at its operating temperature of around 130°C.

Hydrogen carrier gas is supplied at 20 psig to a GC flow controller (Brooks Instruments) set to around 120 ml/min.

The column is interfaced with the atomic absorption burner as shown in Fig. 44 via a $\frac{1}{16}$ in. gas union threaded into a hole in the side of the burner. The outlet inside the burner comes out through four holes in a short manifold to distribute the gas evenly along the flame.

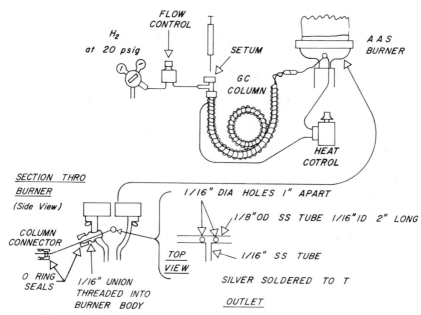

Fig. 45. Gas-chromatographic–atomic absorption apparatus for the determination of lead compounds (178).

The manifold is positioned just under the burner to avoid peak broadening due to mixing with the large gas volume in the burner mixing chamber and to prevent the alkyls from condensing onto the cold mixing chamber walls.

The attachment should be suitable for any instrument. A Varian-Techtron AA5 was used for this work at the following settings: lamp current, 4 mA; wavelength, 2833 Å; slit, 50 μl; lamp height, 0.5 cm; burner, AB41 (10 cm slot); beam height, 10 mm above burner air feed, 15 psig; acetylene feed, 3.5 on flowmeter scale (this corresponds to slightly on the fuel lean side of stoichiometry); mode, absorbance; gain, 6.8; scale expansion, 2×; damping, C.

Recorder, 10 mV, 10 in. chart Honeywell; Electronik 19 with disk integrator; Hamilton 10 μl microsyringe.

Lead standards of 0.40 g Pb/liter were prepared from Octel mix PM 50 (50:50 Pb as TML and TEL) and CR 50 (mixed alkyls) in isooctane.

The lead resonance line at 2833 Å was selected since it gives a much lower noise level and thus a better detection limit than the more sensitive line 2170 Å.

Procedure

The apparatus requires 5–10 min warm-up to stabilize, after which samples and standards can be injected directly into the column at about 5-min intervals. A standard should be run after every six samples at least.

The areas under the peaks are proportional to the lead content of each alkyl and peak areas are best measured using a disk integrator. An electronic digital integrator is not recommended as the oscillations of the signal require too much damping and peak width filtering to allow reliable peak detection and repeatability.

Best results from the disk integrator are obtained by setting the base line just above zero and correcting peak area counts for base line, rather than trying to maintain a zero base line setting.

Determination of Tetraalkyllead Compounds in Biological Samples (179)

Comment on the Method

Tetraalkyllead compounds in fish tissue are selectively extracted using a benzene–aqueous EDTA solution. The benzene extract was evaporated in the presence of nitric acid under a stream of high-purity nitrogen gas at room temperature. Remaining residue was digested with nitric acid at 80–90°C for 2 hr. Selectivity of the method for tetraalkyllead in the presence of other forms of lead, e.g., PbR_3X and PbR_2X_2, was tested. No appreciable quantities of these species were found in the benzene layer. A detection limit of 10 ppb of PbR_4 can be achieved using a 5 g sample.

Reagents and Equipment

A Perkin-Elmer Model 306 atomic absorption spectrometer equipped with deuterium arc background correction was used. Samples were run in the normal furnace mode (argon purge) using the HGA 2100 controller and HGA 74 graphite cell. The temperature program used was as follows: drying, 100°C/40 sec; charring, 500°C/60 sec; atomization 2000°C/5 sec. Samples were analyzed using the 2833 Å-nm resonance line.

Tetraethyllead (99%) and tetramethyllead (68% in toluene) were provided by the Ethyl Corp., Ferndale, Michigan, and were used as working standards at lead concentrations of 10 and 1.0 μg/ml. Inorganic lead standards (in N HNO_3) were prepared from a 1000 μg/ml stock solution of lead nitrate (Fisher atomic absorption standard). Redistilled Fisher scintillation grade benzene was used throughout. Disodium ethylenediaminetetraacetic acid (0.4% w/v, pH 6–7) was made up in glass distilled water. Di- and trisubstituted lead alkyls were obtained from Alfa Ventron Corp., Danvers, Massachusetts.

Procedure

Add benzene (10 ml) and EDTA reagent (10 ml) to a homogenized tissue sample (5 g) in 950 ml glass centrifuge tube fitted with a screw cap, and shake on a Burrell wrist action shaker fitted with extension clamps to give 2.5–3 in. strokes for 10 min. Separate the phases by centrifugation at 2200 rpm for 30 min to give two layers. More discrete separation occurs if the sample is allowed to sit undisturbed overnight after centrifugation.

Transfer an accurately measured 3 ml portion of the benzene layer to a 50 ml calibrated Folin-Wu digestion tube and acidify the contents with 3 ml of concentrated nitric acid (Fisher ACS). Evaporate the benzene layer under a stream of nitrogen (high purity) at room temperature. Digest the residue for at least 2 hr in a heated aluminum block at 80–90°C or until evolution of large amounts of NO_2 ceased. Make the sample up to 10 ml with glass-distilled water and shake with approximately 2 ml of hexane. After removal of the hexane layer, analyze the aqueous phase by the method of standard additions, and calculate the lead concentration following linear regression analysis of peak heights.

Determination of Copper Compounds in Human Serum (177)

Comment on the Method

This procedure can be used to determine copper–EDTA, copper–trien, and copper–histidine in human serum. An anion exchange resin is employed. With this approach the copper EDTA passes through without being held; 2 M ammonium nitrate is used to elute copper trien followed by 4 M nitric acid to elute copper histidine. Because of the low level of copper compound in human serum electrothermal atomization is employed. The effluent from the ion exchange column is collected in the cups of an autosampler.

Equipment

A Perkin-Elmer Model 603 atomic absorption spectrometer was used with a Perkin-Elmer Model 56 recorder and a deuterium arc background correction system. An electrothermal atomizer, the HGA 2100, was used together with a Perkin-Elmer AS-1 auto sampling system. The instrument manufacturers' recommended operating parameters were employed.

Procedure

Use 10 ml burettes as chromatographic columns. Soak Bio-Rex 70, a weakly acidic cation exchange resin in the sodium form, overnight in water and then pack into a column as a slurry in water. Adjust the resin length to be about 8.0 cm when equilibrated in water (the column effluent shows a pH of 6–8).

Apply a mixture of copper–EDTA, copper–trien, and copper–amino acid solutions (25 μl) to the column. Collect the column effluent in plastic sampling cups (capacity about 1.5 ml) and then place in the automatic sampling device to be analyzed by electrothermal atomic absorption. Drying, charring, and ashing temperatures are 100, 700, and 2500°C, respectively. Use the same procedure for the separation of copper–amino acid complexes from human serum (35 ml applied on the column).

Copper–EDTA comes off with a water rinse. Change the solvent to 2 M ammonium nitrate, which elutes copper–trien, followed by 4 M nitric acid, which elutes the copper–amino acid complex.

Determination of Chromium (180)

Comment on the Method

Chromium is reacted with trifluoractylacetone, which forms a volatile compound and can therefore be treated by gas chromatography. The method is excellent for separating chromium from the potential interferences that abound with this element. Better detection limits (1.0 ng) are also obtained compared to the conventional flame-nebulizer technique. Although this is not really a method of chromium speciation, it is included here because of its dependence on gas chromatography.

Reagents and Equipment

This method was developed using a model 303 atomic absorption spectrometer (Perkin-Elmer) equipped with a model 56 recorder. Base line stabilization was accomplished with a model 1021 electronic filter (Spectrum Scientific Corp., Newark, Delaware) with variable cutoff frequencies. The gas chromatograph was fabricated from a hot plate oven (OV10600, Thermolyn Corp., Dubuque, Iowa) by drilling holes in the top and side for injection and exit ports. The injection port was fabricated from $\frac{1}{4}$-in. Swagelok Tee. Nitrogen carrier gas was introduced through a flow meter into the stem of the tee. A septum was placed in the Swagelok nut on one arm and the $\frac{1}{4}$-in.-o.d. Teflon column attached to the opposing arm. At the exit port of the oven, the effluent was carried to the burner of the atomic absorption spectrometer by $\frac{1}{8}$-in.-o.d. copper tubing. The injection port, exit port, and copper tubing were heated by wrapping heating tape around them, and the temperature was controlled by a variable power source (Powerstat, Fisher Scientific). Interface to the spectrometer was accomplished by tapping a threaded hole in the center of the side of the burner heat to accept a $\frac{1}{16}$-in. gold-plated Swagelok union. Both the standard 4 in. single slot and the 2 in. nitrous oxide burner head were successfully used with good results. Preliminary experiments with the triple slot burner head gave a significant decrease in sensitivity. The gas chromatographic column was 24 in. of 3-mm-i.d. Teflon tubing, packed with 0.28 g of 10% SE-30 on Chromosorb WHP 80/100 mesh (Hewlett-Packard, Avondale, Pennsylvania).

Trifluoracetylacetone (Htfa) was obtained from a commercial source (Pierce Chemical Co., Rockford, Illinois), redistilled before use, and stored in a polyethylene bottle at 4°C. Standard solutions of $Cr(tfa)_3$ were prepared by dissolving weighed amounts of the pure chelate in hexane or benzene and diluting volumetrically to the desired concentration. Other chromium chelates were obtained from commercial sources or had been prepared previously.

Solutions of inorganic chromium are prepared by fusion of pure Cr_2O_3 (Alpha Inorganics, Beverley, Massachusetts) with sodium carbonate and dissolution with dilute acid.

Procedure

Digest samples in a reflux apparatus using sulfuric acid and hydrogen peroxide. Place aliquots of the digest (1.00 ml) in reaction tubes fitted with Teflon valves (10 × 100 mm hydrolysis tubes, Kontes Glass, Vineland, New Jersey). Adjust the pH to 5.5–6.0 with a concentrated solution of sodium hydroxide and 0.10 ml of Htfa added. Seal the tubes and place in an oven at 105°C for 2 hr. Remove the tubes from the oven and allow to set at room temperature at least overnight until ready for extraction. Extract the chelate into 0.500 ml of hexane, and inject aliquots of the hexane solution into the GC-AAS for analysis. Up to 0.200 ml injections are used at times for samples of lower concentrations. Inject equal volume aliquots of various concentrations of the standards intermittently with the samples to calibrate the system. Above 0.020 ml, the volume of sample and standard must be equal.

Determination of the Individual Alkyllead Compounds in Street Air (177)

Comment on the Method

The alkyl compounds of lead–tetramethyl, trimethyl ethyl, dimethyl diethyl, methyl triethyl, and tetraethyl—can be determined in levels down to 0.5 ng/m³ for each compound by this method. A 70 liter air sample is taken and the lead compounds trapped in Teflon-lined U tubes containing Chromosorb W at −72°C (dry ice-methanol). The trap is connected to the gas-chromatographic columne and the lead compounds volatilized by heating. Effluent from the gas-chromatographic volume is analyzed for lead using a graphite furnace. The only potential interferences are broad band absorption due to carbon-rich solvent compounds. Solvents can be selected so that this source of error is eliminated.

Reagents and Equipment

The Perkin-Elmer 603 atomic absorption spectrometer used was equipped with a deuterium arc background corrector, an HGA 2100 graphic furnace, and a Perkin-Elmer Model 56 recorder. The radiation source was a Perkin-Elmer electrodeless-discharge lamp operated at 10 W. A Pye gas chromatograph (Series 104) was interfaced to the graphite furnace with a tantalum connector (Fig. 1) machined from a 5.4 mm diameter rod (Ventron Corp.). The glass chromatographic column (150 cm long, 0.6 cm o.d.) was packed with 3% OV-101 on Chromosorb W (80–100 mesh). The effluent was transferred to the furnace by Teflon-lined aluminum tubing (3 mm o.d.), heated electrically to 80°C.

The gas chromatography–mass spectrometry system consisted of a Varian Series 1400 gas chromatograph and a Finnigan quadrupole spectrometer Model 1015-C. The chromatographic column was identical to the above.

Sampling handling was by Hamilton glass syringes of appropriate volumes.

Tetramethyllead and tetraethyllead were obtained from Ventron Corp. Trimethylethyl-, dimethyldiethyl-, and methyltriethyllead were provided by the Ethyl Corp., Ferndale, Michigan. Freshly diluted lead standards were prepared daily in analytical-grade methanol or benzene. The carrier gas for gas chromatography was purified dry nitrogen. The graphite furnace was flushed with purified argon. Analytical-grade nitric acid was used for the leaching of lead from the filters.

The absorption tubes for air samples were U-shaped Teflon-lined aluminum tubes (30 cm long, 3 mm o.d.) packed with 3% OV-101 on

Chromosorb W (80–100 mesh). Moisture was condensed from air by using glass U tubes at −15°C. The air was sampled with a peristaltic pump, Model LG-100, Little Giant Co., Oklahoma City, Oklahoma.

Procedure

Operate the atomic absorption spectrometer with background correction using the 2833 Å lead resonance line and a spectral band width of 7 Å. Heat the graphite furnace continuously for 20 min periods at 1500°C by using the charring stage at the longest time setting. For optimal gas flow from the gas chromatograph, remove the quartz windows from the furnace assemblies. Operate the graphite furnace with an internal flow of 40 ml/min. Because of the high carrier gas flow, the sensitivity is unaffected by the internal gas flow.

Sample air and exhaust directly for organic lead compounds with four parallel traps maintained at −72°C in a dry ice–methanol bath. Precondense water in U tubes at −15°C (sodium chloride in crushed ice). The flow through the traps should average about 70 ml/min. Keep each trap in dry ice until it is attached to the four-way valve installed between the carrier gas inlet and the injection port of the gas chromatograph. Immerse the trap in boiling water. After 1 min, introduce the volatilized fraction of the sample to the gas-chromatographic column by diverting the carrier gas through the four-way valve. Program the gas chromatographic oven for 50–200°C with a heating rate of 40°C/min. Add mixed alkyllead standards to blank traps and run under identical conditions for calibration.

10
Expected New Developments in Atomic Spectroscopy

Atomic absorption, along with other techniques of analytical atomic spectroscopy is undergoing rapid developments in some areas. This chapter is designed to suggest the advances that are worth watching for. Atomic emission and atomic fluorescence spectroscopy are included to give perspective to the anticipated relative roles of the techniques of analytical atomic spectroscopy.

Atomic absorption spectroscopy will continue to be the most commonly used technique in this field. Although not easily adapted to multielement analysis, atomic absorption is relatively rapid and interference free and hence usable by personnel not highly skilled in spectroscopy.

Recent advances in the field of inductively coupled plasma emission spectroscopy and the introduction of commercial instrumentation have led to extravagant claims for the supremacy of this technique. Although inductively coupled plasma emission spectroscopy has multielement capability and is relatively free of chemical interferences, it is expensive and currently appears to require the close attention of a research-trained operator.

Although its initial growth rate has been slow, atomic fluorescence spectroscopy is now close to being a practical method of analysis. This technique possesses many of the advantages of both absorption and emission spectroscopy. Atomic fluorescence instruments are not yet commercially available but the author is particularly optimistic about the future of this analytical technique.

Cutting across the entire discipline of atomic spectroscopy is the development of automated systems. The continuing revolution in microprocessor and mini-computer technology will have a large impact on instrumentation. Initial use of these devices has generally been limited to final data processing, but they will no doubt be applied to total

312

instrument control and signal processing in coming generations of instruments. The first of this type of instrument has already appeared.

It is interesting to view the continuing intense competition between absorption, emission, and fluorescence protagonists. Many of the researchers in these areas view their favorite technique as likely to completely dominate the others, but the author believes that these techniques are largely complementary and each will find an important niche in the extensive analytical field covered by atomic spectroscopy.

RADIATION SOURCES

Atomic absorption spectroscopy still dominates the commercial market. In this application, the hollow-cathode lamp gives adequate output and remains the preferred radiation source for the majority of elements. Difficulties have been encountered with the manufacture of satisfactory hollow-cathode lamps for some refractory and metalloid elements. To overcome these problems and to satisfy the requirement for sources for the rapidly developing field of atomic fluorescence spectroscopy, a great deal of research is being directed toward the development of high-intensity sources. To be useful for atomic absorption, output intensity gains must be achieved without serious line broadening. For atomic fluorescence applications, however, increased spectral output is useful even when accompanied by appreciable line broadening, provided there is no line reversal.

Electrodeless-Discharge Lamps

As judged by commercial availability (181, 182) the most successful type of high-intensity line source is the electrodeless-discharge lamp. This type of lamp has been particularly useful for arsenic, selenium, tin, and phosphorus. Spectral output gains of about a factor of 10 have been achieved while maintaining at least 95% of the absorbance obtainable with hollow-cathode lamps. A factor of 100 useful signal enhancement is commonly obtained in atomic fluorescence applications.

Early in the development of electrodeless-discharge lamps, problems with output instability were profuse. This led to a general disenchantment among analytical chemists who tried to employ these devices in routine analysis. In marked contrast are the commercial lamps, which exhibit a steady output after a 30–60 min warm-up period. This breakthrough was achieved by efficient coupling of the excitation energy to the atom source in the lamp bulb. This results in an even distribution of heat throughout the bulb and stable temperatures.

Unfortunately, it is not yet possible to manufacture satisfactory electrodeless-discharge lamps for most refractory elements. This remains a very active area of research.

High-Intensity Lamp

Another source of intense spectral radiation is the "boosted" or high-intensity hollow-cathode lamp developed about ten years ago by Sullivan and Walsh (183). These lamps gave intensity increases of 10–100-fold compared to typical hollow-cathode lamps available at that time. However, further improvements in the design of hollow-cathode lamps reduced these gains to the extent that the higher cost of production for high-intensity lamps was not warranted. Although one company, Varian-Techtron, did produce high-intensity lamps for a short period about 10 years ago, they are no longer commercially available.

Recent work, particularly by Lowe (184) on an improved design of a sealed high-intensity lamp and by Lowe (185) and Sullivan (186) on demountable high-intensity lamps, has again stimulated interest in this approach. Preliminary results show that compared to modern hollow-cathode lamps, signal enhancements of about a factor of 10 can be obtained without significant line broadening. In the improved design of the sealed lamp, the boosting discharge passes through the center of the hollow cathode, resulting in more efficient excitation of the sputtered atom cloud. Lamps of this type have been used to advantage by Larkins (187) in nondispersive fluorescence.

Intensity gains with demountable high-intensity lamps derive largely from their operation at the optimum gas pressure (approximately 2 Torr argon) and from water cooling of the cathode, which allows the use of higher current densities. The continuous flow of inert gas flushes away impurities, which results in rapid stabilization of intensity and a discharge of high spectral purity.

When these lamps are used with a recently developed automated gas control unit (188) the change from one element to another and stabilization of the lamp output can be accomplished in 2–3 min.

Other Radiation Sources

Continuum sources are of interest in dispersive atomic fluorescence spectroscopy and for use in conjunction with resonance spectrometers in atomic absorption. The types most commonly used in spectroscopy are the EIMAC* 150 W or 500 W high-pressure xenon arc lamps, which incorporate an internal parabolic reflector. When these sources are used in conjunction with resonance detectors, the effective line width is controlled by the absorption profile of the detector. The uv-enhanced version of the EIMAC xenon arc has useful intensity throughout the visible and uv regions.

Tunable dye lasers have recently been produced that extend down to about 2500 Å and with linewidths comparable to those produced by atomic spectral-line sources. Many of these sources have sufficient intensity to cause saturation of excited states. Under this condition, fluorescence intensity is no longer proportional to source intensity. The dye laser, to date, has been used almost exclusively for investigations of atomic and molecular fluorescence processes (189). Analytical application awaits a decrease in price (presently more than $15,000), improvements in ease of operation, and ability to produce wavelengths below 2500 Å.

If it could be made sufficiently small and low cost, the tunable laser could eventually replace most other radiation sources for use with atomic spectroscopy equipment.

ATOM RESERVOIRS

Flames remain the atomizers of choice for the broadest range of applications in absorption and fluorescence spectroscopy. Developments in electrothermal and cathodic sputtering atomizers have made these devices more attractive in the applications to which they apply. On the other hand, the inductively coupled plasma has now been developed to the stage where it is beginning to displace flames, arcs, and sparks as the favored atomizers for the widest range of applications in emission spectroscopy. The Grimm low-pressure glow discharge is finding increasing favor in the field of alloy analysis.

Flames

There is a continuing research emphasis on flames, which in large part involves investigations of their chemistry and physics. Of particular interest to analytical spectroscopy are investigations of light scattering, molecular absorption, and molecular fluorescence, and in this regard studies by Alkemade and co-workers (e.g., 190) should be consulted.

* EIMAC, Division of Varian, San Carlos, California.

Surprisingly, there has been little recent effort to improve nebulizer systems and nebulization processes with flame atomizers. There is, however, a continuing interest in the nebulization of solids. Willis (191) found that solids with particle sizes smaller than 325 mesh could be nebulized without clogging problems. Accuracies acceptable for geochemical prospecting purposes were achieved.

The problem of nebulization of small volumes has received some attention. Manning (192) introduced reproducible volumes using small sample cups attached directly to the capillary and measured the signal peak height.

No new flames of particular analytical interest have been proposed recently. However, of great interest was the report by Walsh (193) outlining the use of a flame as a resonance spectrometer for atomic absorption and emission spectroscopy. This is discussed in detail in the section on spectrometers.

Air–acetylene is still the most commonly used flame for atomic absorption determinations and is now being used increasingly in atomic fluorescence work. However, hydrogen-based flames generally yield better detection limits in fluorescence and will continue to be used in cases where it is established that they will not result in interferences due to matrix effects or scatter.

The problem of scatter of incident radiation by unevaporated solvent or solid particles in the flame is a major limitation to the use of atomic fluorescence for the determination of trace levels of elements in solutions containing high concentrations of dissolved solids. Most correction methods proposed to date are based on the measurement of scatter using an auxiliary lamp (194,195) but a recent proposal (196) to use the Zeeman effect deserves further investigation. The main limitation to this latter approach would seem to be the reduction in light intensity caused by the associated use of a polarizer, especially at low wavelengths.

Electrothermal Atomizers

Prior to the introduction of electrothermal atomizer instrumentation, atomic absorption could be used by the unskilled operator with very few problems. In contrast, work with electrothermal atomizers, which is often carried out at concentrations near the detection limit, is fraught with many error-prone operations including sample injection, selection of proper ashing and atomizing times and temperatures, and especially alignment and proper operation of background correction equipment.

These problems are further complicated by deficiencies in all but the most recent equipment. In view of these problems it is not surprising to find such a large number of research papers devoted to investigations of these atomizers.

A recent important development in this field is the realization by commercial manufacturers of the importance of temperature control to the proper operation of electrothermal atomizers. For minimization of interferences and optimization of atomization, close temperature control during ramp and static modes is essential. Until recently, it has been normal practice in commercial devices to apply a regulated voltage to the furnace and assume that this will give a sufficiently constant and reproducible temperature. This approach is not satisfactory because of variations in the resistance of furnace elements, due particularly to aging. Some new instruments are capable of close regulation of furnace temperature using feedback control. In this regard, the addition of microprocessor programmable control would be a significant advance.

In addition to poor temperature control, matrix problems, molecular absorption, and light scattering plague work with electrothermal atomizers. Detection limits obtained with electrothermal atomizers can be seriously degraded when analyses are performed in the presence of complex inorganic matrices. To obviate this problem the technique of selective volatilization of matrix before analyte or vice versa has been proposed. Unfortunately, present commercial equipment cannot perform this separation successfully for a wide variety of samples. Equipment with close temperature control provides better operating characteristics and should improve this type of separation.

Several authors, e.g., Katskov et al. (197), Robinson and Wolcott (198), Van Loon and Radziuk (199), and Koop et al. (200) have suggested physically separating the volatilization and atomization functions of the electrothermal atomizer. This allows better time resolution of matrix and analyte absorption signals.

In the work described by Koop et al. (200) the sample is volatilized using a conventional electrothermal atomizer and the vapors are injected into a flame atomizer. Using this approach the detection limit obtained for lead in complex matrices was only degraded by a factor of 10 compared to that obtainable on soft water samples using the electrothermal atomizer in the normal way. In addition, nonspecific absorption effects were found to be negligible for all samples.

Separation of volatilization and atomization can yield other benefits. Van Loon and Radziuk (199), Koop et al. (200), and Robinson and Wolcott (198) showed that this approach can be used for metal speciation studies—a topic of much current interest to be discussed below. Katskov *et al.* (197) used their equipment for the analysis of solid samples.

In spite of significant reductions in nonspecific absorption achieved through design changes to electrothermal atomizers, background correction is still of crucial importance. The most common approach to this correction is the use of a continuum source. Experience suggests that the uncritical use of this device can result in errors due to over- or undercorrection. For reliable results the line source and continuum source beams must be adjusted to follow coincident paths through the electrothermal atomizer. In addition, the two beams should fill the same fraction of the aperture of the monochromator. This latter condition is difficult to achieve because of differences in the physical dimensions of the glow from these two different types of lamp.

Of most recent interest is the work involving use of the Zeeman effect for background correction (196). For this technique the spectral line source (a specially designed electrodeless-discharge or hollow-cathode lamp) is placed in a strong magnetic field. With suitable transitions this results in splitting of the emission line into its π and $\pm\sigma$ components and if the magnetic field is sufficiently intense the $\pm\sigma$ components will be shifted outside the absorption line. As the π and σ components are polarized at right angles to each other, they are readily separated so that the absorption at each component can be measured separately. Recently an instrument based on Zeeman modulation (201) was introduced commercially. In this equipment the magnetic field is applied around the furnace atomizer.

Another serious source of error in work with electrothermal atomizers is the sample introduction technique. This procedure is usually carried out by micropipet but autosamplers have been recently introduced with greatly improved reproducibility of sample introduction.

Solid-Sample Analysis

The goal of the analytical chemist is to be able to analyze solid samples directly. Decompositions are time-consuming and error prone. Unfortunately, only a little progress has been made to date in the analysis of complex solids directly by atomic absorption.

Several researchers, e.g., Willis (202), have nebulized solids, suspended in a stream of liquids, into a flame. When a particle size less than 325 mesh is used, clogging does not occur. This approach was used for sulfides and stream sediments.

L'vov (203) pioneered the determination of solids directly using electrothermal atomization. In his approach the solid is placed in a large graphite tube. A smaller graphite tube is inside the above tube. Upon electrothermal heating, the vapors from the sample travel through the walls of the internal tube into the optical beam. A similar approach is used with a flame as an atomizer. In this latter case the superiority of a flame as an atomizer is utilized to reduce nonspecific interferences.

Attempts to analyze most geological solids directly in a conventional electrothermal device have not been successful. The nature and quantity of the matrix make background correction impossible. Success using this approach has been obtained for a limited range of organic samples. The latter are organic materials very low in high-boiling-point inorganic salts.

Recently in our laboratory (200) we have used a commercial furnace interfaced with a flame for the analysis of solids. Because atomization is accomplished in a flame, it is possible to eliminate nonspecific background interference or bring it within tolerable limits. Enhanced detection limits compared to conventional flame analysis are achieved because the furnace pulses all the analyte into the atomizer over a very short interval.

In conclusion, although progress with direct solid analysis by atomic absorption has been slow to date, recent developments give hope of a brighter future.

Low-Pressure Glow Discharge

Recently, there has been a notable increase in interest in the use of the low-pressure gas discharge in both atomic absorption and emission investigations. Of particular note in this field is the work by Laqua and co-workers (204), Butler and co-workers (205), and Gough (206). In atomic absorption studies, the low-pressure gas discharge is used simply as a method of sampling from a solid conducting surface while in emission studies the discharge is used both to produce free atoms from the sample and also to excite these atoms to produce emission. This method of sampling by cathodic sputtering is also of great potential interest for use in conjunction with atomic fluorescence. Nonmetallic samples can also be analyzed if they are first finely ground and then pressed into a pellet with copper or silver powder. Since the sample is atomized in a low pressure of inert gas this technique should be useful for the determination of important nonmetallic elements such as carbon, sulfur, phosphorus, and nitrogen.

An important observation in cathodic sputtering was reported in a paper by Gough (206). These workers observed persistent background absorption, which they assigned to aggregates of atoms formed subsequent to the sputtering process. They were able to correct for this problem by an electronic method based on the difference in the rates with which the free atoms and the agglomerates diffuse out of the absorption light path.

Butler has developed a simple instrument using a combination of a low-pressure gas discharge emission lamp and a demountable resonance spectrometer (low-pressure gas discharge fluorescence). This instrument has proven to be very useful for application on the ferrous metallurgical industry.

INDUCTIVELY COUPLED PLASMA

The commercial introduction of emission quantometers using inductively coupled plasma sources has had a major impact on analytical atomic spectroscopy. The two crucial components of such a unit are the source and the spectrometer. To date, the spectrometers used have been those available from conventional arc/spark-type instrumentation.

Some controversy exists over the desirability of using high- or low-power inductively coupled plasmas (207,208). Presently most commercial instrumentation is of the low-power type. These plasmas have power ratings of 2–5 kW and electron temperatures of about 8000 K.

Inductively coupled plasmas have a very simple argon gas emission spectrum, which is in marked contrast to the complexity of the band spectra produced in flames. The chemical interference is believed by most workers to be minimal in inductively coupled plasmas because of the very high temperature. However, because of this high temperature, plasmas produce a large number of excited state and ion lines. When elements such as iron and calcium are present in high concentration, emission intensities can be so high that light scattering in the spectrometer can be a serious source of error. This latter problem will be discussed further below.

Other types of plasma source (e.g., the DC and microwave plasmas) have been proposed over the years. The microwave plasma is particularly useful for the determination of metals that can be introduced in the gas phase, e.g., mercury and arsenic.

SPECTROMETERS

New developments in this area have been rather sparse. Over the past 30 years the mechanically ruled grating has slowly replaced the prism as the dispersive element in most monochromators. No significant changes are likely to occur in monochromators in the near future and the most likely development in this area will be the slow displacement of mechanically ruled gratings by holographically produced ones, provided the latter can be made at a lower cost than that of replication.

For some applications in atomic spectroscopy the grating monochromator can now be replaced by flame- or sputtering-type resonance spectrometers.

The only other developments in the area of wavelength isolation are gradual improvements in interference filters and a continuing low level of interest in Echelle monochromators. Neither of these are likely to be of much general impact.

Holographic Gratings

Holographic gratings are manufactured by exposing a photoresist layer, on a suitable substrate, to the interference pattern produced by two monochromatic laser beams, followed by suitable development and reflective coating processes. The principal virtues of these gratings are their freedom from "ghosting" associated with periodic errors in the ruling engine drive screws and a very much lower level of stray light, the latter a consequence of a smoother line profile. For gratings produced on modern ruling engines both of these defects are already at insignificantly low levels for most applications. Another advantage claimed for the holographic production technique is the ability to form gratings on a curved surface. This is very difficult using mechanical ruling techniques.

Because of their low stray light levels, holographic gratings are supposed to provide a significant improvement over mechanically ruled ones for use in conjunction with intense emission sources such as inductively coupled plasmas. As stated earlier, they are likely to find more general use if their production cost can be made comparable to that of replication.

Resonance Spectrometers

If an atomic vapor of some element is illuminated with radiation that includes wavelengths corresponding to resonance lines of that element, then the atoms will absorb some of the radiation and produce fluorescence. The intensity of the fluorescence will be proportional to the intensity of the incident radiation over the absorption linewidth of the atoms and thus provide a means of effectively isolating the resonance lines of the element. An atom cloud used in this fashion to isolate atomic resonance lines is termed a resonance spectrometer.

The concept of the resonance spectrometer was first suggested by Russell and Walsh (209) in the early years of atomic absorption. The proposal was to use sputtered metal vapors as spectrometers for atomic absorption spectroscopy. Because of their compact form these spectrometers held the promise of a practical multielement approach to atomic absorption. Unfortunately, commercial instrument manufacturers did not find the concept attractive. Reasons for this attitude included the need for a separate sealed detector for each element and the relatively poor S/N ratio of early instrumentation.

However, two recent developments portend a brightening future for these devices. First, Walsh (210) introduced the concept of using an atomic vapor produced in a flame as a resonance spectrometer, and Larkins and Walsh (211) have shown that this technique is applicable to both absorption and emission spectroscopy. Second, demountable sputtering-type resonance spectrometers have now become practical. Butler (212) in a recent report described the use of this type of resonance spectrometer in his work with low-pressure, glow discharge emission spectroscopy.

DETECTORS

The most interesting development in the area of light detection has been the application of TV-type multichannel detectors to spectroscopy. A good summary of the various types available and their performance characteristics has been given recently by Talmi (213,214). The main use for these detectors is in multielement analysis but they have also been useful in investigations of the distribution of atoms in flames (215).

In general, the principal disadvantages of these detectors are associated with their poor sensitivity in the UV (216), a limited dynamic range, and high cost. While prices can be expected to fall, those detectors most likely to become cheaper are the ones based on silicon technology. Should prices fall as low as predicted these devices would find much greater application in spectroscopy despite their other disadvantages.

The standard type of photomultiplier has changed little since it was first introduced. The main improvement has been the development of low-noise, broad-response cathodes which allow coverage of the 200–900 nm range by one tube.

At the other extreme are the limited response (160–320 nm) solar-blind photomultipliers. These devices, useful particularly in nondispersive systems (217), should also be helpful in reducing the effects of stray light when working with a monochromator in the range around 200 nm.

INSTRUMENTS

Emphasis in the foregoing has been on components of analytical spectroscopy instrumentation. There are several instrumental developments that should be briefly highlighted.

Inductively Coupled Plasma Emission Spectrometry

To date the emphasis in commercial inductively coupled plasma emission spectrometry has been on the quantometer. Unfortunately, while maximizing ease of operation this approach leads to problems resulting from direct spectral overlap and stray light interferences when used in laboratories that must analyze a wide range of different sample types.

A number of workers, particularly those with a strong background in arc, spark, and flame emission spectroscopy, have adopted a more conservative approach. These researchers employ polychromator with photographic spectral recording or a scanning monochromator for the identification of interferences together with the direct reader.

Inductively coupled plasma emission spectroscopy is a very important development in analytical atomic spectroscopy. At present, the technique is undergoing a particularly difficult teething period. Therefore, to date, unlike in the case of atomic absorption spectroscopy, a highly skilled research spectroscopist is essential for the reliable, routine operation of this equipment. In the future, the use of Echelle gratings for better resolution or holographic gratings for better reduction of stray light and/or the use of background correction may result in improved performance of inductively coupled plasma emission quantometers. It is unlikely that these developments will obviate the need for the attention of a research spectroscopist.

Multielement Atomic Fluorescence Spectroscopy

Since the demise of the Technicon AFS-6 there has been no commercial atomic fluorescence spectroscopy equipment on the market. It appears that the time is right for a new introduction. The technique has distinct advantages over both atomic emission and absorption spectroscopy. A good deal of research has been done to overcome the most important source of error with this technique—light scatter. Larkins (218) has designed and built a relatively inexpensive seven-channel nondispersive instrument with automatic background correction. The system has been used successfully for the analysis of waters, sediments, and alloys.

Atomic Absorption Instrumentation

There is an ever-increasing interest in the automation of equipment. Microprocessors are presently available at prices that allow their inclusion in relatively inexpensive commercially available atomic absorption equipment. At this writing, microprocessors are in general use in the output section of equipment. Several microprocessor-controlled instruments have now appeared.

Minicomputers are increasing in capability while decreasing in cost. Soon most laboratories with large data output requirements will employ minicomputers, probably on a time-shared basis, with most of their electronic equipment.

These very important steps forward must not be taken without caution. There is a tendency for laboratory personnel to view automation as a guarantee of error-free analyses. It is important to realize that the same chemical and physical principles apply to manual and automated chemical analysis. Automation can have the adverse effect of so divorcing the laboratory technician from the actual analysis that he fails to take account of potential error sources. Automation must be accompanied by a redoubled effort to ensure a proper knowledge of all the chemistry and spectroscopy involved in an analysis.

Atomic absorption has continually defied attempts to produce a commercially successful simultaneous multielement (more than two elements) instrument. Reasons include serious optical constraints and relatively shorter linear dynamic working range (about 10 times less than emission). Both Instrumentation Laboratories Ltd. and Fisher-Jarrel Ash Co. Ltd. have offered two element atomic absorption units for some time. Recently, Perkin-Elmer Corp. introduced the Model 5000, which is a six-element sequentially operated device. This instrument is completely microprocessor controlled and can be programmed either by keyboard or magnetic card to automatically analyze six elements, one after the other in samples in a carousel. A number of simultaneous multielement atomic absorption units have been proposed. None has become commercially available.

References

1. K. Müller, *Z. Phys.* **65,** 739 (1930).
2. A. Walsh, *Spectrochim. Acta* **7,** 108 (1955).
3. W. Grotian, "Graphische Darstellung der Spectren Von Atomen und Ionen mit ein zwei und drei Valenzelektronen." Springer-Verlag, Berlin and New York, 1928.
4. A. G. Mitchell and M. W. Zemansky, "Resonance Radiation and Excited Atoms." Cambridge Univ. Press, London and New York, 1971.
5. G. Herzberg, "Atomic Spectra and Atomic Structure." Dover, New York, 1944.
6. L. de Galan and G. F. Samaey, *Spectrochim. Acta, Part B* **24,** 679 (1970).
7. A. Walsh, private communication.
8. E. A. Boling, *Spectrochim. Acta* **22,** 425 (1966).
9. K. Fuwa and B. L. Valee, *Pap., 1965 East. Anal. Symp., 1965.*
10. "Analytical Methods for Atomic Absorption Spectroscopy." Perkin-Elmer, Norwalk, Connecticut, 1976.
11. H. T. Delves, *Analyst* **95,** 431 (1970).
12. C. W. Fuller, "Electrothermal Atomization for Atomic Absorption Spectroscopy," Anal. Monogr. Ser. No. 4. Chem. Soc., London, 1977.
13. B. V. L'vov, *Inzh.-Fiz. Zh.* **2,** 44 (1959).
14. H. Brandenburger and H. Bader, *At. Absorpt. Newsl.* **6,** 101 (1976).
15. T. S. West and X. K. Williams, *Anal. Chim. Acta* **45,** 27 (1969).
16. J. W. Robinson, D. K. Wolcott, and L. Rhodes, *Anal. Chim. Acta* **78,** 285 (1975).
16a. D. C. Manning and R. D. Ediger, *At. Absorpt. Newsl.* **15,** 42 (1976).
16b. D. S. Gough, P. Hannaford, and A. Walsh, *Spectrochim. Acta, Part B* **28,** 197 (1973).
17. W. R. Hatch and W. L. Ott, *Anal. Chem.* **40,** 2085 (1968).
18. Y. Kimura and V. L. Miller, *Anal. Chim. Acta* **27,** 325 (1962).
19. W. Holak, *Anal. Chem.* **41,** 1712 (1969).
20. J. C. Van Loon, *At. Absorpt. Newsl.* **11,** 685 (1972).
21. J. D. Norris and T. S. West, *Anal. Chem.* **45,** 1423 (1974).
21a. C. W. Frank, W. G. Schrenk, and C. E. Meloan, *Anal. Chem.* **38,** 1005 (1966).
22. N. Omenetto and G. Rossi, *Anal. Chim. Acta* **40,** 195 (1968).
23. R. D. Ediger, *At. Absorpt. Newsl.* **14,** 127 (1975).
24. C. Hendrix-Jongerius and L. de Galan, *Anal. Chim. Acta* **87,** 259 (1976).
24a. H. Koizumi and K. Yasuda, *Spectrochim. Acta, Part B* **31,** 236 (1976).
25. L. R. P. Butler, *Anal. Chem. Explor. Min. Process. Mater., 1976.*
26. J. C. Van Loon, *Proc. Int. Conf. Heavy Met. Environ. 1970* p. 349 (1975).
27. J. F. Uthe, F. A. J. Armstrong, and K. C. Tam, *J. Assoc. Off. Anal. Chem.* **54,** 866 (1971).
28. R. Lauwreys, J. P. Buchet, H. Roels, A. Berlin, and J. Smeets, *Clin. Chem.* **21,** 551 (1975).
29. B. Loescher, "Report to Participants of Toronto Lead Intercomparison Study." Toronto, 1975.
30. G. Kirkbright and H. N. Johnson, *Talanta* **20,** 433 (1973).
31. B. Fleet, K. V. Liberty, and T. S. West, *Talanta* **17,** 203 (1970).
32. C. H. Williams, D. J. David, and O. Iisma, *J. Agric. Sci.* **59,** 381 (1962).
33. L. Barnes, Jr., *Anal. Chem.* **38,** 1083 (1966).
34. K. C. Thompson and D. R. Tomerson, *Analyst* **99,** 595 (1974).
35. J. C. Van Loon and E. J. Brooker, *Anal. Lett.* **7,** 505 (1974).
36. A. E. Smith, *Analyst* **100,** 300 (1974).
37. L. R. P. Butler and E. Norval, *S. Afr. Med. J.* p. 2617 (1974).

38. P. Bailey and T. A. Kilroe-Smith, *Anal. Chim. Acta* **77**, 29, 1975.
39. C. Ida, T. Tanaka, and K. Yamasaki, *Bunseki Kagaku* **15**, 1100 (1966).
40. D. C. Voyve and H. Zeitlin, *Anal. Chim. Acta* **69**, 27 (1974).
41. T. C. Rains, private communication.
42. W. W. Vaughn and J. H. McCarthy, Jr., *U.S., Geol. Surv., Prof. Pap.* **501**, D123 (1964).
43. A. C. West, V. A. Fassel, and R. N. Knisely, *Anal. Chem.* **45**, 1586 (1973).
44. L. L. Sundberg, *Anal. Chem.* **45**, 1460 (1973).
45. A. Strasheim and G. J. Wessels, *Appl. Spectrosc.* **17**, 65 (1963).
46. T. W. Steele, J. Levin, and J. Capelowitz, *Natl. Inst. Metall., Repub. S. Afr., Rep.* **1696** (1975).
47. R. Lockyer and G. E. Hames, *Analyst* **84**, 385 (1959).
48. A. C. Menzies, *Anal. Chem.* **32**, 898 (1960).
49. A. E. Pitts, F. E. Beamish, and J. C. Van Loon, *Anal. Chim. Acta* **50**, 181 (1970).
50. J. C. Van Loon, Z. *Anal. Chem.* **246**, 122 (1969).
51. M. M. Schnepfe and F. S. Grimaldi, *Talanta* **16**, 1461 (1969).
52. F. S. Grimaldi and M. M. Schnepfe, *Talanta* **17**, 617 (1970).
53. J. C. Van Loon, *At. Absorpt. Newsl.* **8**, 6 (1969).
54. J. G. Sen Gupta, *Miner. Sci. Eng.* **5**, 207 (1973).
55. R. G. Mallett, D. C. G. Pearton, and E. J. Ring, *Natl. Inst. Metall., Repub. S. Afr., Rep.* **1086** (1970).
56. A. Jansen and F. Umland, Z. *Anal. Chem.* **251**, 101 (1970).
57. R. C. Mallett and R. L. Breckenridge, *Natl. Inst. Metall., Repub. S. Afr., Rep.* **1318** (1971).
58. A. E. Pitts and F. E. Beamish, *Anal. Chim. Acta* **52**, 405 (1970).
59. D. C. G. Pearton and R. C. Mallett, *Natl. Inst. Metall., Repub. S. Afr., Pap.* **1595** (1974).
60. G. L. Everett, *Analyst* **101**, 348 (1976).
61. I. Rubeska, *Can. J. Spectrosc.* **20**, 156 (1975).
62. J. B. Willis, *Appl. Opt.* **7**, 1295 (1968).
63. R. Goecke, *Talanta* **15**, 871 (1968).
64. S. Sachdev, J. W. Robinson, and P. W. West, *Anal. Chim. Acta* **37**, 12 (1967).
65. J. A. F. Gidley and J. T. Jones, *Analyst* **85**, 240 (1960).
66. R. J. Gibbs, *Nature (London)* **180**, 71 (1973).
67. J. Stary, "The Solvent Extraction of Metal Chelates." Macmillan, New York, 1964.
68. G. H. Morrison and H. Freiser, "Solvent Extractions in Analytical Chemistry." Wiley, New York, 1957.
69. Y. A. Zolotov, "Extraction of Chelate Compounds." Ann Arbor-Humphrey Sci. Publ., Ann Arbor, Michigan, 1970.
70. J. O. Kinrade and J. C. Van Loon, *Anal. Chem.* **46**, 1894 (1974).
71. E. Lakanen, *At. Absorpt. Newsl.* **5**, 17 (1966).
72. W. F. Fitzgerald, W. B. Lyons, and C. D. Hunt, *Anal. Chem.* **46**, 1883 (1974).
73. J. P. Riley and D. Taylor, *Anal. Chim. Acta* **40**, 479 (1968).
74. F. J. Fernandez, *At. Absorpt. Newsl.* **12**, 93 (1973).
75. J. C. Van Loon, Electrothermal Atomization, Internal Laboratory Procedures, 1977.
76. H. W. Fairbairn, *U.S., Geol. Surv., Bull.* **980**, 1–71 (1951). (A cooperative investigation of precision and accuracy in chemical, spectrochemical and modal analysis of rocks.)
77. J. A. Corbett, W. C. Godbeer, and N. C. Watson, *Proc. Australas. Inst. Min. Metall.* **250**, 51 (1974).
78. M. D. Amos and J. B. Willis, *Spectrochim. Acta* **22**, 1325 (1966).

79. S. Abbey, *Geol. Surv. Can., Prof. Pap.* **74-19** (1974).
80. R. Goquel, *Spectrochim. Acta, Part B* **26,** 313, 1971.
81. M. S. Cresser and R. Hargitt, *Anal. Chim. Acta* **82,** 203 (1976).
82. R. Sanzolone and T. T. Chao, *Anal. Chim. Acta* **86,** 163 (1976).
83. L. P. Greenland and E. Y. Campbell, *Anal. Chim. Acta* **87,** 323 (1976).
84. F. J. Langmyhr, R. Solberg, and Y. Thomassen, *Anal. Chim. Acta* **92,** 105 (1977).
85. F. J. Langmyhr and Y. Thomassen, *Z. Anal. Chem.* **264,** 122 (1973).
86. S. H. Omang and P. E. Paus, *Anal. Chim. Acta* **56,** 393 (1971).
87. A. Armannson, *Anal. Chim. Acta* **88,** 89 (1977).
88. J. C. Van Loon, Rocks and Minerals, Current Laboratory Practice, 1978.
89. H. Agemian and A. S. Y. Chau, *Anal. Chim. Acta* **80,** 61 (1975).
90. P. Boar, personal communication.
91. J. Murphy, *At. Absorpt. Newsl.* **14,** 151 (1976).
92. C. Block, *Anal. Chim. Acta* **80,** 369 (1975).
93. W. L. Gladfelter and D. W. Dickenhoof, *Fuel* **55,** 360 (1976).
94. M. Tomljanovic and Z. Grobenski, *At. Absorpt. Newsl.* **14,** 52 (1975).
95. F. J. Langmyhr and P. E. Paus, *Anal. Chim. Acta* **43,** 397 (1968).
96. M. D. Amos and P. E. Thomas, *Anal. Chim. Acta* **32,** 139 (1965).
97. B. M. Gatehouse and J. B. Willis, *Spectrochim. Acta* **17,** 710 (1971).
98. R. G. Smith, J. C. Van Loon, J. R. Knechtel, J. L. Fraser, A. E. Pitts, and A. E. Hodges, *Anal. Chim. Acta* **93,** 61 (1977).
99. J. J. Lynch and G. Mihailou, *Geol. Surv. Can., Prof. Pap.* **63-8** (1963).
100. J. N. Bishop, L. A. Taylor, and P. L. Diosady, *Minist. Environ. Ont., Can. Rep., March, 1975.*
101. J. C. Van Loon, Exploration Geochemistry Methods, Current Lab Practice, 1978.
102. J. C. Van Loon, J. Lichwa, D. Ruttan, and J. Kinrade, *Air, Water, Soil Pollut.* **2,** 473 (1973).
103. F. E. Beamish, "Analytical Chemistry of the Noble Metals." Pergamon, Oxford, 1966.
104. E. E. Bugbee, "A Textbook of Fire Assaying," 3rd ed. Wiley, New York, 1940.
105. J. C. Van Loon, *Z. Anal. Chem.* **246,** 122 (1969).
106. T. W. Oslinsky and N. H. Knight, *Appl. Spectrosc.* **22,** 530 (1968).
107. M. M. Kruger and R. V. D. Robert, *Natl. Inst. Metall., Repub. S. Afr., Rep.* **1432** (1972).
108. J. G. Tweed, personal communication.
109. A. A. Vasilyeva, I. G. Yedelevich, L. M. Gindim, T. V. Lanbina, R. S. Shulmann, I. L. Kotlarevsky, and V. N. Andrewsky, *Talanta* **22,** 745 (1975).
110. P. E. Moloughney and G. H. Faye, *Talanta* **23,** 377 (1976).
111. R. E. Theirs, *Methods Biochem. Anal.* **5,** 273 (1957).
112. G. D. Martinie and A. A. Schilt, *Anal. Chem.* **48,** 70 (1976).
113. W. I. Adrian, *At. Absorpt. Newsl.* **10,** 96 (1971).
114. W. I. Adrian, personal communication.
115. T. J. Ganje and A. L. Page, *At. Absorpt. Newsl.* **13,** 131 (1974).
116. A. J. Jackson, L. M. Mitchell, and H. J. Schmachen, *Anal. Chem.* **44,** 1064 (1972).
117. S. B. Cross and E. C. Parkinson, *Interface* **3** (1974); *At. Absorpt. Newsl.* **13,** 107 (1974).
118. J. C. Van Loon, Internal Lab Method for Organic Samples, 1977.
119. P. Schramel, *Anal. Chim. Acta* **67,** 69 (1977).
120. J. C. Van Loon and R. B. Cruz, *Anal. Chim. Acta* **72,** 231 (1974).
121. F. J. Fernandez, *Clin. Chem.* **21,** 558 (1975).

122. E. Berman, *At. Absorpt. Newsl.* **3,** 111 (1964).
123. L. E. Kopito, M. A. Davis, and H. Schwachman, *Clin. Chem.* **20,** 205 (1974).
124. F. J. Fernandez and H. L. Kahn, *At. Absorpt. Newsl.* **10,** 1 (1971).
125. A. H. Jones, *Anal. Chem.* **48,** 1472 (1976).
126. H. Zachariasen, I. Andersen, C. Kostol, and R. Barton, *Clin. Chem.* **21,** 562 (1975).
127. D. Littlejohn, G. S. Fell, and J. M. Ottaway, *Clin. Chem.* **22,** 1719 (1976).
128. F. J. Langmyhr and D. L. Tsalev, *Anal. Chim. Acta* **92,** 79 (1977).
129. J. P. Cali, G. N. Bowers, Jr., and D. S. Young, *Clin. Chem.* **19,** 1208 (1973).
130. T. C. Rains and O. Menis, *J. Assoc. Off. Anal. Chem.* **55,** 1339 (1972).
131. M. A. Evenson and C. T. Anderson, Jr., *Clin. Chem.* **21,** 537 (1975).
132. K. J. Julshamn and O. R. Braekkan, *At. Absorpt. Newsl.* **14,** 49 (1975).
133. S. Henning and J. L. Jackson, *At. Absorpt. Newsl.* **12,** 100 (1973).
134. E. E. Cary and A. H. Allaway, *J. Assoc. Off. Anal. Chem.* **58,** 433 (1975).
135. W. Holak, *At. Absorpt. Newsl.* **12,** 63 (1973).
136. S. Slavin, G. E. Peterson, and P. C. Lindahl, *At. Absorpt. Newsl.* **14,** 57 (1975).
137. F. J. Langmyhr and J. Aomodt, *Anal. Chim. Acta* **87,** 483 (1976).
138. M. M. Morris, M. A. Clarke, V. W. Tripp, and F. G. Carpenter, *J. Agric. Food Chem.* **24,** 45 (1976).
139. J. A. McHard, J. D. Winefordner, and S. V. Ting, *J. Agric. Food Chem.* **24,** 1950 (1967).
140. R. A. Baetz and C. T. Kenner, *J. Agric. Food Chem.* **23,** 41 (1975).
141. P. L. Larkins and J. B. Willis, *Spectrochim. Acta, Part B* **29,** 319 (1974).
142. D. R. Thomerson and W. J. Price, *Analyst* **96,** 825 (1971).
143. D. B. Radcliff, C. S. Byford, and P. B. Osman, *Anal. Chim.* **75,** 427 (1975).
144. K. E. Burke, *Analyst* **97,** 19 (1972).
145. M. Kirk, E. G. Perry, and J. M. Arritt, *Anal. Chim. Acta* **80,** 163 (1975).
146. W. B. Barnett and E. A. McLaughlin, *Anal. Chim. Acta* **80,** 285 (1975).
147. F. Shaw and J. M. Ottaway, *At. Absorpt. Newsl.* **13,** 77 (1974).
148. F. F. Bell, *Anal. Chem.* **45,** 2296 (1973).
149. M. Bedard and J. D. Kerbyson, *Can. J. Spectrosc.* **21,** 64 (1976).
150. R. M. Hamner, D. L. Lechak, and R. Greenberg, *At. Absorpt. Newsl.* **15,** 122 (1976).
151. G. G. Welcher, O. H. Kreige, and J. Y. Marks, *Anal. Chem.* **46,** 1227 (1974).
152. T. R. Dulski and R. R. Bixler, *Anal. Chim. Acta* **91,** 199 (1977).
153. P. N. Vijan, J. A. Pimenta, and A. C. Rayner, "Air Quality Laboratory Procedure." Ontario Ministry of the Environment, Toronto, Canada.
154. L. J. Purdue, E. R. Erione, and R. J. Thompson, *Anal. Chem.* **45,** 527 (1973).
155. P. N. Vijan and G. R. Wood, *At. Absorpt. Newsl.* **13,** 33 (1974).
156. A. Zdrojewski, L. Bubois, and N. Quickert, *Sci. Total Environ.* **6,** 165 (1976).
157. Y. Thomassen, R. Solberg, and J. E. Hanssen, *Anal. Chim. Acta* **90,** 279 (1977).
158. R. D. Reeves, C. J. Molnar, M. T. Glen, J. R. Ahlström, and J. D. Winefordner, *Anal. Chem.* **44,** 2205 (1972).
159. C. J. Molnar, R. D. Reeves, J. D. Winefordner, M. T. Glen, J. R. Ahlström, and J. Savoy, *Appl. Spectrosc.* **26,** 606 (1972).
160. C. Saba and K. J. Einsentraut, *Anal. Chem.* **49,** 454 (1977).
161. S. H. Omang, *Anal. Chim. Acta* **56,** 470 (1971).
162. M. P. Bratzel, Jr. and C. L. Chakrabarti, *Anal. Chim. Acta* **61,** 25 (1972).
163. Y. E. Arakting, C. L. Chakrabarti, and I. S. Maines, *Spectrosc. Lett.* **7,** 97 (1974).
164. J. T. Oleiko, *J. Am. Oil Chem. Soc.* **53,** 480 (1976).
165. E. L. Henn, *At. Absorpt. Newsl.* **12,** 109 (1973).
166. F. J. Langmyhr and J. T. Hakedal, *Anal. Chim. Acta* **83,** 127 (1976).

167. C. W. Fuller and J. Whitehead, *Anal. Chim. Acta* **68**, 407 (1974).
168. R. D. Thompson and T. J. Hoffman, *J. Pharm. Sci.* **64**, 1863 (1975).
169. J. C. Van Loon, J. Lichwa, D. Ruta, and J. D. Kinrade, *Air, Water, Soil Pollut.* **12**, 473 (1973).
170. J. J. Labrecque, *Appl. Spectrosc.* **30**, 625 (1976).
171. B. Kolb, G. Kemnner, F. H. Schlesser, and E. Wiedeking, *Z. Anal. Chem.* **221**, 166 (1966).
172. S. E. Manahan and D. R. Jones, *Anal. Lett.* **6**, 745 (1973).
173. F. J. Fernandez, *At. Absorpt. Newsl.* **16**, 33 (1977).
174. J. C. Van Loon and B. Radziuk, *Can J. Spectrosc.* **21**, 46 (1976).
175. J. C. Van Loon and B. Radziuk, *Sci. Total Environ.* **5**, 190 (1976).
176. J. G. Gonzalez and R. T. Ross, *Anal. Lett.* **5**, 683 (1972).
177. J. C. Van Loon, Chemical Speciation Procedures, Current Laboratory Practice, 1977.
178. D. T. Coker, *Anal. Chem.* **47**, 386 (1975).
179. G. R. Soirota and J. F. Uthe, *Anal. Chem.* **49**, 423 (1977).
180. W. R. Wolf, *Anal. Chem.* **48**, 1717 (1976).
181. W. B. Barnett, *At. Absorpt. Newsl.* **12**, 142 (1973).
182. Canadian Westinghouse Co. Ltd., 777 Walkers Line, Burlington, Ontario, Canada.
183. J. V. Sullivan and A. Walsh, *Spectrochim. Acta* **21**, 721 (1965).
184. R. M. Lowe, *Spectrochim. Acta, Part B* **26**, 201 (1971).
185. R. M. Lowe, *Spectrochim. Acta, Part B* **31**, 257 (1976); *Int. Conf. At. Spectrosc., 5th, 1975* Abstr. B. 15 (1975).
186. J. V. Sullivan, *Abstr. Pap.—Int. Conf. At. Spectrosc., 5th, 1975* Abstr. B. 13 (1975).
187. P. L. Larkins, *Spectrochim. Acta, Part B* **26**, 477 (1971).
188. P. L. Larkins, to be published.
189. N. Omenetto, P. Benetti, L. P. Hart, J. D. Winefordner, and C. T. J. Alkemade, *Spectrochim. Acta, Part B* **28**, 289 (1973).
190. C. Th. J. Alkemade, H. P. Lijnse, and T. J. M. J. Vierbergen, *Spectrochim. Acta, Part B* **27**, 149 (1972).
191. J. B. Willis, *Anal. Chem.* **47**, 1752 (1975).
192. D. C. Manning, *At. Absorpt. Newsl.* **14**, 99 (1975).
193. A. Walsh, *Analyst* **100**, 764 (1975).
194. T. C. Rains, M. S. Epstein, and O. Menis, *Anal. Chem.* **46**, 207 (1974).
195. P. L. Larkins and J. B. Willis, *Spectrochim. Acta, Part B* **29**, 319 (1974).
196. H. Koizumi and K. Yasuda, *Spectrochim. Acta, Part B* **31**, 237 (1976).
197. D. A. Katskov, L. P. Kruglinkova, B. V. L'Vov, and L. K. Polzik, *Zh. Prikl. Spektrosk.* **20**, 739 (1974).
198. J. W. Robinson and D. K. Wolcott, *Anal. Chim. Acta,* **74**, 43 (1975).
199. J. C. Van Loon and B. Radziuk, *Can J. Spectrosc.* **21**, 46 (1976).
200. D. J. Koop, M. D. Silvester, and J. C. Van Loon, *Abstr. Pap., 26th Pittsburgh Conf.* Abstr. No. 383 (1975).
201. H. Koizumi and K. Yasuda, *Abstr. Pap., 28th Pittsburgh Conf.* Abstr. No. 208 (1977).
202. J. B. Willis, *Anal. Chem.* **47**, 1752 (1975).
203. B. V. L'vov, "Atomic Absorption Spectral Analysis," 1st ed. Nauka, Moscow, 1966.
204. K. Laqua, *5th International Conf. on Atomic Spectroscopy* Abstr. No. 22, Melbourne, Australia, Aug. 25–29 (1975).
205. L. R. P. Butler, K. Kroeger, and C. D. West, *Spectrochim. Acta, Part B* **30**, 12 (1975).
206. D. S. Gough, *Anal. Chem.* **48**, 1926 (1976).
207. V. A. Fassel and R. N. Kniseley, *Anal. Chem.* **46**, 1110A and 1155A (1974).
208. S. Greenfield, McD. H. McGeachin, and P. B. Smith, *Anal. Chim. Acta* **84**, 67 (1976).

209. B. J. Russell and A. Walsh, *Spectrochim. Acta* **15,** 883 (1959).
210. A. Walsh, *Int. Conf. At. Spectrosc., 5th, 1975* President's Address (1975).
211. P. L. Larkins and A. Walsh, *Int. Conf. Heavy Met. Environ., Proc., 1975* P249 (1975).
212. L. R. P. Butler, *Abstr. Pap.—Int. Conf. At. Spectrosc. 5th, 1975* Abstr. No. B12 (1975).
213. Y. Talmi, *Anal. Chem.* **47,** 658A (1975).
214. Y. Talmi, *Anal. Chem.* **47,** June 697A (1975).
215. T. E. Edmonds and G. Horlick, *Abstr. Pap., 28th Pittsburgh Conf.* Abstr. No. 455 (1977).
216. N. G. Howell and G. H. Morrison, *3rd FACS Meet.* Abstr. No. 59 (1976).
217. P. L. Larkins, R. M. Lowe, J. V. Sullivan, and A. Walsh, *Spectrochim. Acta, Part B* **24,** 187 (1969).
218. P. L. Larkins, *3rd FACS Meet.* Abstr. No. 281 (1976).

Index